The Energy Equation

Requirements, Resources, and Projections

by

Aniruddha B. Joshi

ISBN:148409834X
ISBN-13:978-1484098349

DEDICATION

This book is dedicated to my wife, parents, and daughters. It would
have been impossible to write this book without them.

CONTENTS

ACKNOWLEDGMENTS

I am grateful to the United Nations for making a number of statistical databases available online. I also thank BP p.l.c., London, for making the yearly energy statistics available online. Various departments of the US government and its agencies have also made a wealth of information available online. The solar insolation data, available from NASA for the entire world, is a particularly useful resource.

x

1. INTRODUCTION

It was in the month of June in the year 2000 that San Jose and other towns in the Silicon Valley in California suffered from rolling blackouts for the first time in many years. The dot-com boom had brought in tremendous prosperity to the region. Economy was booming and so was the demand for energy. The uncomfortably low difference between the available electric generation capacity and the demand for electricity created an energy crisis. The energy crisis continued for several months, though blackouts were mostly avoided by purchasing power at astronomical prices in the spot market. It was only after the dot-com bubble burst that the energy situation normalized. This crisis drove the Pacific Gas and Electric Company (PG&E) to bankruptcy and exposed manipulations by Enron, which led to its eventual fall.

On the other side of the globe, in India, the state of Maharashtra always boasted of surplus power, while most states in India suffered serious shortages. In the spring of 2005, something changed. The Indian economy had been growing at a healthy rate following the economic liberalization. Maharashtra being the most industrialized state, had greatly benefited from the economic boom. A direct result of the booming economy was the growth in energy consumption, particularly the rise in the demand for electricity. At the same time, most of the generation, transmission, and distribution of electricity was controlled by the state government, which still embraced socialist policies. For several years, hardly any, new generation capacity had been added to the grid. A direct result of this situation was the gap in the demand and supply of electricity. The resultant energy crisis caused rolling blackouts in the entire state of Maharashtra, except Mumbai, the state capital. The rolling blackouts implied that villages suffered daily blackouts of up to 16 hours! Even the information technology centers, such as the city of Pune, suffered from daily blackouts of up to 4 hours. In spite of assurances by politicians that the situation would be normalized in a couple of years, the rolling blackouts continue to affect about 90 million residents even today, in the year 2012.

The astronomical rise in the price of crude oil in the year 2008 and its subsequent fall impacted the global economy. Not only did the price of crude oil rise, it also coincided with the rise in the price of natural gas, coal, precious metals, and other commodities. This price rise preceded the recession that impacted almost all the developed economies in the world. Even in the year 2012, nations in the Euro zone are suffering from a challenging economic situation, while the US economy is still limping. Economic situation in the rapidly developing nations such as India continues to be worrisome while the price of crude oil is back over $100 per barrel.

We live in interesting times. On one hand, there are challenges of human development, with billions still living in poverty. The rapidly developing Asian economies, mainly India and China, have lifted hundreds of millions from poverty but many more still aspire to get access to the basis necessities. At the same time, the demand for energy from these developing economies is rising rapidly. The rising demand has led to a worrying rise in the price of crude oil and coal. It appears that the supply of crude oil is being stretched to a limit where no-more increase in supply may be possible. Developed nations are still in an economic slowdown that is likely to affect investments in the development of clean and renewable sources of energy, while dependence on fossil fuels is leading to rising levels of carbon dioxide in the atmosphere, which is expected to have a disastrous impact on the global climate in the future.

This book is an attempt to introduce the world of energy to the reader. The book also attempts to explore the linkage of energy consumption with economics and human development. An attempt is also made to make projections for the future demand for energy and an analysis of the capability of sources of energy to meet this demand. Any reader with an inquisitive mind and high school science education should be able to grasp the contents of this book. This book is organized in the following manner:

The second chapter in this book covers some basic concepts and terminology related to the world of energy, that will prepare the reader for the subsequent topics.

Chapter 3 introduces the linkage between energy usage, economy, and human development. This linkage makes it possible to estimate the energy requirements in the future.

Chapter 4 is about fossil fuels. Can fossil fuels support the energy demand from developing economies with growing population? This chapter attempts to project the future demand for energy and evaluates if the reserves of fossil fuels can satisfy the demand. An attempt is also made to evaluate if it will be possible of transition from crude oil to other fossil fuels.

Chapter 5 is about the renewable sources of energy. While these sources continue to hold promise, at present they satisfy only a small portion of the total demand for energy. This chapter analyses various issues in the widespread adoption of renewable sources of energy.

Chapter 6 discusses nuclear energy. While some claim that nuclear energy is undergoing renaissance, the disastrous accident at Fukushima, Japan, has cast fresh doubts about the future for nuclear energy. At the same time, claims are made that nuclear energy is indispensable for the development of economy of India, and perhaps the most ambitious program for harnessing nuclear energy is currently underway in India. This chapter analyzes the pros and cons in the adoption of the nuclear energy option.

Chapter 7 attempts to take a peek at the future, based on the projected availability of energy resources.

Glossary and References are provided at the end of this book. The reader is encouraged to look them up whenever necessary. The references are identified within chapter text using square brackets, for instance, [BP 2010].

Most of the analysis and projections in this book are based upon statistical data available in the public domain. United Nations (UN) database provides some very useful data to link energy consumption, economics, and human development. This book also uses data from the US Energy Information Administration (EIA) and the BP Statistical Review of World Energy from BP p.l.c., London, UK. In many cases, the energy scenario in India is used for analysis and projections, while in some cases USA is used as an example. Yet, this book is not just about India or USA. The analysis and projections will be similar for most large nations in the world that use diverse sources of energy.

This book should help the reader to gain an understanding of the world of energy resources and their impact on economy and development. Such an understanding should be useful in taking decisions in personal life which relate to the use of energy in any way. The decisions could range from the purchase of a new car to buying solar panels to install on the rooftop. An understanding of the world of energy resources may also help in making investment decisions and to prepare for the future. While this book does not attempt to predict the future, certain future scenarios will be discussed in detail. They should help prepare for the future, based on individual needs, goals, and preferences.

Finally, this book is not about global warming or peak-oil. There are a number of books and websites on these topics which take an extreme view of the future, predicting an end of the industrial civilization. The reader will notice that some websites even offer extreme solutions to survive the demise of civilization. These solutions range from innocuous ones such as growing food in the backyard, to extreme ones such as hoarding gold and guns! In this book, such an extreme view will not be taken. Data and science will be the guiding forces. While the end of industrial civilization is an extreme possibility, there are other optimistic scenarios as well. Human endeavor should be given its due credit! Perhaps the civilization is at the crossroads, where making the right choices may in fact make life more fulfilling, exciting, and healthy!

2. ENERGY BASICS

Energy and Civilization

It won't be an exaggeration to say that practically every aspect of our civilization is dependent on the modern sources of energy. Centuries ago, the only source of energy available to man was the heat obtained by burning firewood and coal. In that era, the sole quest in life for most people was to work to fulfill the basic survival needs of the individual and the society. Such a society would measure very poor on the scale of human development. Literacy, Life Expectancy, and Infant Mortality of that society would be much worse compared to any developing nation today. It was the development of science and technology that unlocked the sources of energy which provided the fuel for development of civilization to the complex form that exists today. Development of energy sources and methods of energy conversion allowed a rapid rise in energy consumption as well as in the standard of living and human development. It is no coincidence that countries that rank the highest on the scale of human development also have some of the highest levels of energy consumption.

The unprecedented swings and volatility in the price of crude oil since 2007 have coincided with the turmoil in the financial markets. There are concerns about the dwindling fossil fuel resources and claims have been made that crude oil production is fast approaching a peak and there is an impending permanent decline. At the same time, several renewable resources are taking prominence. Nuclear energy seems to have gotten a new lease of life along with the protests against it. It would not be wrong to say that energy, and for that matter civilization, is at the crossroads. A fall in the energy available for consumption is certain to impact the civilization. This impact could range from the collapse of civilization in an extreme case, to a smooth transition to new sources of energy in the best-case. In this book, an attempt will be made to objectively examine the complex subject of energy resources based on the publicly available data. A comparison of renewable energy sources will also be presented. Based on the data and calculations, it will be possible to imagine what the energy future has in store for the civilization.

Before starting with analysis and projections for various energy sources, it will be worthwhile to review some basic concepts. Some of these concepts are generally a part of any high school science curriculum. In the remaining part of this chapter, concepts useful for a study in the field of energy will be reviewed.

Various Forms of Energy

The forms of energy useful in our day to day lives are:
- Electrical Energy
- Heat Energy
- Mechanical Energy
- Chemical Energy

Each of these forms of energy fulfill some of our needs. Development of science and technology has made it possible to convert energy in one form to another. For instance, chemical energy stored in petrol or diesel is converted into mechanical energy in the internal combustion engine of a car. A ceiling fan converts electrical energy into mechanical energy, while a kitchen stove converts chemical energy stored in LPG/CNG into heat. It is these energy conversions that practically run our civilization. James Watt's steam engine was the first efficient, practicable, and commercially successful implementation of conversion from chemical energy to mechanical energy. Since then, a number of energy conversion techniques were devised that resulted in the development of devices that range from an Internal Combustion Engine (ICE) and steam turbines to fuel cells. Another important aspect has been the use of electricity and the associated devices that range from electric motors and generators, to LED lights. The use of electricity makes is very simple to transport energy over a large geographic area and the use of energy can take place thousands of kilometers away from its source. The devices used to convert energy from one form to another have significantly improved productivity, standard of living, and in general, have significantly contributed to human development. It will be no exaggeration to say that practically all aspects of civilization, whether it is agriculture, industry, transportation, commerce, entertainment, education, research, or household

Illustration 2.1: Steam Turbine: Produced by Siemens AG, Germany, picture by Christian Kuhna, Siemens AG, Germany

chores, are dependent upon using some source of energy and the associated energy conversions.

Energy Sources

Firewood, coal, and biomass in general, are the sources of energy that have been in use for centuries. To a very small extent, even windmills were also used as a source of energy in some parts of the world. As a part of the industrial revolution, other sources of energy were developed, as science and technology progressed. Coal was the most commonly used energy source until the early part of the 20th century. After that, energy use transitioned to crude oil as the most popular source of energy. In most of the developed nations, coal is no longer used for household applications, and its use is limited to electricity generation and the smelting of metals. In the last 50 years, the use of natural gas has been rising, though it will be too early to say that a transition is taking place from crude oil to natural gas. Natural gas has replaced the use of coal in domestic and industrial heating applications as well as for the generation of electricity in some countries. Yet, in many countries, coal continues to dominate as a preferred source of energy. Hydroelectricity and nuclear energy have also been the other important sources of energy. Hydroelectricity was developed rapidly over the past 100 years. In fact, in developed countries, most feasible and economical sources have already been harnessed. Most of the new hydroelectric projects are coming up in the developing nations. The golden era for nuclear energy was probably the 3rd quarter of the previous century, 1950 to 1975. In that era, there were dreams that nuclear energy would be so abundant that it would be a waste to meter it! Though these dreams were never realized, nuclear energy remains an important source of energy. In the last 20 years, renewable sources of energy are getting prominence. Use of wind energy has increased and the use of solar power for electricity as well as for water heating applications has been rising. Now that the threat of global warming is better understood, there is a significant pressure to increase the use of renewable sources of energy. Yet, as things stand today, fossil fuels, which are, coal, petroleum, and natural gas; continue to supply a majority-part of the energy that mankind uses. In 2009, fossil fuels were used for 80% of the energy that the world consumed [IEA 2011].

Units for Measuring Energy

Lord Kelvin famously said, "If you can't measure it, you can't improve it". To understand the subject of energy, it is first necessary to understand the units in which energy is commonly measured. The units used for measuring energy depend on the application and the type of energy. The units may also be different in different countries. The most common unit used to measure electrical energy is kilowatt-hours (kWh). Our monthly electric bill generally mentions this as the number of *units* consumed in the month. Energy in the form of heat is measured in calories. Heat energy contained in food is measured in kilo-calories, which is also denoted as Calories, with an upper case 'C'. It is also common to use the unit BTU (British Thermal Units) to measure the energy delivered as heat. In some countries, natural gas consumed for household heating/cooking is measured in Therms. Another unit of energy used in the petroleum and energy industry is TOE, or a Tonne of Oil Equivalent. It is the amount of energy contained in a tonne of crude oil.

The SI unit of energy is Joules. Though Joules is used by most of the scientific community, for the day to day applications, all the above units of energy are still in use. These units can be converted into one another; for instance, 1 calorie corresponds to 4.186 Joules and 1 kWh corresponds to 3.6 million joules (MJ). For the purpose of this book, Joules is used as the basic unit to measure energy. Other units of energy are also used in a few cases, when the use is most convenient and logical. Table 2.1 summarizes some useful conversion ratios between the various units of energy.

1 calorie	4.186 Joules
1 Calorie	4186 Joules
1 kWh	3,600,000 Joules (3.6 MJ)
1 Therm (USA)	105,480,400 Joules (105.48MJ)
1 BTU	1,055 Joules
1 TOE (Tonne Oil Equivalent)	42,000,000,000 Joules (42 GJ)

Table 2.1: Energy Unit Conversions

A Joule is a very small unit of energy. A 40 Watt fluorescent tube light consumes 40 Joules of energy in just 1 second. To measure the energy usage in a country, we obviously need a larger unit. Table 2.2 summarizes the units derived from a Joule to measure the energy use on a larger scale.

Unit	Abbreviation	Conversion
Mega Joules	MJ	1 MJ = 1,000,000 Joules
Giga Joules	GJ	1 GJ = 1,000,000,000 Joules
Tera Joules	TJ	1 TJ = 1,000,000,000,000 Joules
Exa Joules	EJ	1 EJ = 1,000,000,000,000,000,000 Joules

Table 2.2: Energy Units

Examples of usage of these units:
- Energy consumed by a household electric water heater in 1 hour = 10.8 MJ.
- Per capita yearly energy consumption in India = 18 GJ.
- Total yearly energy consumption in India = 22 EJ.

Difference Between Power and Energy

Power is the amount of energy delivered in unit time. The commonly used unit for measuring power is Watts (W). One Watt is the power required to deliver 1 Joule of energy in 1 second. A typical LCD television set has a power rating of about 200 Watts. This power rating implies that the television set will consume about 200 Joules of energy in one second. A typical household water heater has a power rating of 3,000 Watts, implying the consumption of 3,000 Joules of energy every second. Using the water heater continuously for 1 hour results in the consumption of 10.8 MJ of energy, which is the same as 3 kWh. Power rating of electrical appliances are normally specified in Watts or kilowatts (kW), while the power capacity of motors and engines is usually specified in horsepower (HP). One HP equals about 746 Watts. Typical power ratings for some of the commonly used household appliances are summarized in Table 2.3.

Appliance	Power Rating
Compact Fluorescent Lamp (CFL)	5 W to 40 W
Incandescent Bulb	15 W to 200 W
Television	50 W to 500 W
Refrigerator	100 W to 400 W
Microwave Oven	900 W to 2 kW
Electric Water Heater	1 kW to 3 kW
Air Conditioner	1 kW to 3 kW

Table 2.3: Power Rating of Household Appliances

Compare this information with engine power of cars and motorcycles:

Vehicle	Power Rating
Honda Activa (India)	6 kW
Bajaj Pulsar (India)	15 kW
Tata Nano (India)	26 kW
Maruti/Suzuki Alto	35 kW
Honda City	73 kW
Toyota Corolla	118 kW
BMW 740	200 kW to 400 kW

Table 2.4: Power Rating of Some Vehicles

Energy Content of Fuels

The energy content of a fuel is the amount of heat energy obtained as a result of combustion of the fuel. This heat is either used directly to heat some substance or transformed into mechanical energy to perform some useful work. Energy content of a fuel is measured in terms of energy in million Joules, obtained per kg of the fuel. Table 2.5 summarizes the energy content of some important fuels. The energy content of a fuel is also known as energy density of the fuel.

Fuel	Energy Content (MJ/kg)
Coal (lignite)	15 to 19
Coal (Bituminous/Anthracite)	27 to 30
Petrol (Gasoline)	47
Diesel	45
Kerosene	46
Ethanol (Alcohol)	29
LPG	49
CNG	54
Wood	15
Hydrogen gas	123
Biodiesel	37 to 42
Charcoal	30

Table 2.5: Energy Content of Fuels

The numbers in this table are approximate. Impurities and ash content in coal, for instance, can change the actual energy content. It can be seen that all the petroleum products have an excellent energy density, making them very suitable fuels. Hydrogen has the highest energy density, but it is not found naturally in a molecular form (H_2), that is necessary for its use as a fuel.

Energy Efficiency

When we use a gas stove for cooking, only a certain percentage of the heat generated due to the combustion of the gas is actually utilized to heat food. The remaining heat is lost in the surroundings. On a larger scale, in a coal based thermal power station, only about 38% of the energy obtained by combustion of coal is converted into electricity. The remaining 62% is lost as heat. An internal combustion engine in a car burns petrol to convert the resultant heat into mechanical energy that drives the car. Only a portion of the heat energy is converted into mechanical energy. Energy efficiency specifies the percentage of the source energy converted into the desired form. For any energy conversion device, efficiency is never 100%. Conversion from heat to mechanical energy tends to have poor efficiency, while conversion from electrical energy to mechanical energy is normally quite efficient. When studying energy sources, especially the renewable ones, it is important to evaluate the energy efficiency.

The concept of energy efficiency not only applies to power stations but also to devices in everyday use, ranging from light bulbs to air conditioners, and water pumps to television sets. Higher the efficiency of a device, lesser is the energy use. In a

Lighting technology is undergoing a transformation with LED lights becoming cheaper. Energy efficiency of lighting has increased by almost an order of magnitude over the incandescent lights. Light output, or luminous flux in technical terms, is measured in Lumens. The incandescent light delivers about 12 to 15 Lumens per Watt of electrical power. The Compact Fluorescent Lamps (CFL) deliver about 60 to 70 Lumens per Watt, while LED lights that deliver 130 Lumens per Watt are available. LED lights with even higher efficiency are under development. The LED lights also have a very long operational life. Cree Inc., USA, specifies a lifetime of 35,000 hours for XLamp XP-E LEDs [Cree]. These LEDs can last for 20 years and more! It is only a matter of time when LED lighting will replace the other types of lighting in most applications to save a significant amount of energy.

growing economy like India, electricity shortages are very widespread. Using devices with high energy efficiency results in savings that directly translate to lesser electricity demand. The Bureau of Energy Efficiency in India tests various household appliances for energy efficiency under the Standards and Labeling Program [BEE]. The appliances are rated on a 1 to 5 star scale, with 5 being the rating for the most energy efficient appliance. The bureau even specifies building codes for designing energy efficient buildings. It is expected that a similar rating system will also be in place for cars very soon.

An even larger energy efficiency program has been in operation in USA. This is the ENERGY STAR program, operated jointly by the US Environment Protection Agency and the US Department of Energy. In its 2010 report about achievements, it was

Illustration 2.2: Household LED Lighting in Pune, India

claimed that this program saved 170 million metric tonnes of GHG emissions in 2010, equivalent to the annual emissions from 33 million vehicles [ENERGY STAR 2010]. This information illustrates how important it is to improve energy efficiency. Energy saved is energy generated!

Primary Energy: An Important Concept

The energy sources that are found in nature, and which are not derived as a result of some energy conversion, are the primary sources of energy. All fossil fuels; viz. crude oil, coal, natural gas, are the primary sources of energy. Similarly, nuclear energy and solar energy in its various forms, are primary sources of energy as well. Primary sources of energy, is a very important concept. In any country, energy consumption takes place in various forms. Electricity is the most convenient form of energy for many applications. A major part of the electricity is produced by burning fossil fuels. Hydroelectricity, nuclear energy, and electricity derived using direct solar conversions, such as photovoltaic, is termed as primary electricity. Strictly speaking, energy conversion does take place even for primary electricity. For instance, energy contained in electromagnetic radiation from sun is converted into electricity in photovoltaic generation.

To measure the total energy usage by a nation (or by an economy), it is common to use primary energy as the quantity for measurement. When fossil fuels are used as the energy source, primary energy is calculated by multiplying the quantity of fossil fuels used by their energy content. The energy content of various fuels is specified in Table 2.5. For the consumption of primary electricity, the corresponding primary energy is obtained by multiplying the consumed energy by a factor that corresponds to the efficiency of thermal power stations. In other words, primary energy that corresponds to primary electricity is calculated as if thermal power stations were utilized to generate the same amount of electricity. For instance, the *BP Statistical Review of World Energy*, published yearly by BP p.l.c. in UK, assumes an efficiency of 38% for thermal power stations. Hence, primary electrical energy is multiplied by 2.63 to obtain the corresponding primary energy.

Primary energy is sometimes measured using MTOE (Million Tonnes Oil Equivalent), as the unit for measurement. In this book, both, Joules and MTOE will be used to measure primary energy. Primary energy consumption is a useful benchmark to measure development of a country, economic activity, as well the energy efficiency of an economy. These linkages will be analyzed in the next chapter.

Household Energy Use

Before taking a look at the energy usage by a country, it will be useful to take a look at the household use of energy. In this book, two scenarios of energy usage, from two diametrically opposite sides of the world, will be presented. One is from Pune, India, for a middle class family of four and the other for the same family while they were in Cupertino, California, in USA, with approximately the same standard of living. The examples provided here are just instructive samples. They will provide the reader some insight into how to measure the household energy usage. The energy usage has been measured over 1 year so that seasonal changes in energy consumption are accounted for. While measuring gasoline (petrol) consumption, the use of public transport facilities is not accounted for and only the direct energy purchases are counted. Both cities, Pune, and Cupertino, have a poor public transport, and a majority of the middle class does not use public transport.

Household energy consumption generally takes place in 3 forms:

- Electricity
- CNG/LPG for cooking, water heating, and room heating where applicable
- Petrol (Gasoline) for personal vehicles

Apart from this, practically everything that a household consumes corresponds to some energy consumed somewhere. Such consumption is not accounted for in this example and only direct energy consumption is calculated. It can be seen that different units are used for measuring the consumption of electricity, petrol, and CNG/LPG. These units are converted into GJ to calculate the primary energy consumption.

Pune, India	Electricity	Cooking Gas (LPG)	Petrol	Notes
Consumption in basic units	3,000 kWh/year	85 kg/year	470 liters/year	
Consumption in GJ	10.8 GJ/year	4.2 GJ/year	16 GJ/year	
Equivalent primary energy consumption GJ	28.4 GJ/year	4.2 GJ/year	16 GJ/year	Assumes 38% efficiency of thermal power plants

Table 2.6: Household Primary Energy Consumption in Pune, India

Table 2.6 calculates the household primary energy consumption in Pune to be 48.6 GJ/year. The next table calculates the household primary energy consumption for the same family in Cupertino, CA, USA. The units used in USA for measuring CNG consumption is Therms.

Cupertino, CA, USA	Electricity	Natural Gas (CNG)	Petrol	Notes
Consumption in basic units	3,000 kWh/year	437 Therms/year	1100 liter/year	
Consumption in GJ	10.8 GJ/year	46.1 GJ/year	37.6 GJ/year	1 Therm = 105.48 MJ
Equivalent primary energy consumption GJ	28.4 GJ/year	46.1 GJ/year	37.6 GJ/year	Assumes 38% efficiency of thermal power plants

Table 2.7: Primary Energy Consumption for a Cupertino, California family

The above table calculates the household primary energy consumption in Cupertino to be 112 GJ/year.

There are several reasons why the consumption in USA is more than double of that in India. We will not explore the reasons here but it is sufficient to mention that reasons are related to lifestyle as well as climate. The consumption numbers should provide some indication about the growth in domestic energy consumption that will take place in India as the middle class grows.

Total Primary Energy Consumption in a Country

It is possible to calculate the Total Primary Energy (TPE) consumption for a country in a way similar to the household energy consumption. Energy consumption in a country is based on the various primary sources of energy:

- Fossil Fuels: Coal, Petroleum, Natural Gas, and associated liquids
- Nuclear Energy
- Hydro-electricity
- Biomass
- Other Renewable Sources of Energy (Photovoltaics, Wind etc.)

Various information sources are available on the internet that report the TPE consumption of various countries. The consumption in the form of biomass may be ignored in the calculations as it is quite difficult to get reliable numbers for the biomass consumption. Energy Information Administration (EIA) of the US Department of Energy measures the TPE consumption in a country using Quadrillion BTU as the unit. The yearly *Statistical Review of World Energy* by BP p.l.c. UK, measures the TPE consumption using MTOE as the unit. The BP report only counts the commercially traded fuels and primary electricity generated using nuclear, hydro, and modern renewable energy sources for the TPE calculation. According to this report, published in June 2011, TPE consumption in USA and India in 2010 was the following:

Country	Oil	Natural Gas	Coal	Nuclear	Hydro	Renew-able	Total (MTOE)	Total (EJ)
USA	850	621	524.6	192.2	58.8	39.1	2285.7	96
India	155.5	55.7	277.6	5.2	25.2	5	524.2	22

Table 2.8: TPE Consumption, USA and India

The TPE consumption is converted to units EJ (Exa Joules) in the last column. It can be seen that USA consumes more than 4 times the energy as compared to India while having only about one third the population!

Power Density and Energy Density

This topic is not about the energy content of a fuel. It is about the land used by a source of energy. Every energy source requires some resources on the land to generate energy. For instance, a coal based thermal power station requires linkage to a coal mine for the supply of coal. A hydroelectric power station requires a reservoir that spans a large area. Power density and energy density are two important concepts that provide information about the land use by a given energy source. Power density is the capacity of a square meter of land to provide power on a continuous basis. Energy density is the amount of energy output available from a square meter of land (used to tap an energy source) in one year. Power density is measured in Watts/m^2 and energy density is measured in kWh/m^2/year. Energy density can be obtained by multiplying the power density by 8.76 since the number of hours in a year are 8760 and energy density is measured in kWh as against Wh.

The power and energy density tend to be quite low for renewable energy sources. As a result, they need a large land area to provide useful energy supply. The power density for nuclear energy sources tends be very high due to the high concentration of energy per kg of fuel (natural Uranium). Energy density for conventional fossil fuels tends to be high. The density for non-conventional fossil fuels such as oil shale tends to be low.

Power and energy density provide a good means to compare energy sources. In this book, these parameters will be used to mainly compare the renewable sources of energy.

Capacity Factor of Power Station

The concept of capacity factor applies to power generation by an electric power station. Capacity factor is also know as the Plant Load Factor (PLF). Any power station that generates electricity is rated to generate a certain amount of power. For instance, units 3 and 4 of the Tarapur Atomic Power Station near Mumbai, India, are rated to generate 540 MW each. This does not mean that the power station will actually generate the rated power continuously for a year. There will be times when the power station will need to be shutdown for various reasons. There will also be times when the power station does not operate at the rated power. The capacity factor essentially measures the capacity of the power station to operate at the rated power on a continuous basis. Thus, if the Tarapur unit operated for 320 days in a year at an average power of 500 MW then the capacity factor is calculated to be:

$$capacity\ factor = \frac{(320 \times 500)}{(365 \times 540)} \times 100$$

Which is about 81%. In other words, the capacity factor is the capacity of the power station to generate energy with respect to the maximum power rating.

Different power sources have different capacity factors. Nuclear and coal based thermal power stations normally operate at a high capacity factor. Capacity factor for wind power is dependent upon the availability of wind. Wind turbines operate at about 20% capacity factor for most locations [MEDA 2009]. Hydroelectric power stations can work at a high capacity factor as well as at a low capacity factor, depending on the requirement and the availability of water in the reservoir. Natural gas based thermal power stations are quite flexible too. Capacity factor is an important measure to compare the various methods of power generation since it measures the ability to generate energy.

Dispatchable Power Sources

Apart from capacity factor, dispatchability is another important attribute of a source of power. This attribute measures the ability of the power source to adapt the generation of electricity to demands from the electric grid. The demand from electric grid is not constant. It varies during the course of the day and there are seasonal variations as well. Not all sources of electricity are dispatchable. For instance, the availability of wind energy is totally dependent on the presence of wind. Similar argument applies for solar power. Even for nuclear power and coal based thermal power, generation cannot be started or stopped very quickly and the magnitude of power generated cannot adapt quickly to the demands from the grid. Such sources of electricity are considered non-dispatchable. Whereas, for hydroelectric power, it is much easier to start or stop the generation. It is also possible to adapt the amount of electricity generated to the demands from the grid. Such sources are dispatchable sources of power. In general, the non-dispatchable sources of electricity with a high capacity factor are used to supply the base load on the grid. Base load is the load that is always present on the grid, irrespective of the time of the day or the season. The nuclear power plants and coal-based thermal power plants have a high capacity factor though they are not dispatchable. They are the ideal candidates to supply the base load. The demand peaks on the grid are handled by dispatchable sources of electricity. One of the challenges in the large scale use of solar power and wind power is their characteristic of low capacity factor and non-dispatchability.

The Concept of Entropy and its Importance

The concept of entropy originates from the second law of thermodynamics. Energy tends to flow from a concentrated form to a dispersed form. For instance, petrol (gasoline) is an example of energy stored in a concentrated form. When an automobile engine burns petrol, the energy stored in petrol is released and gets dispersed in various forms. The energy is partly dispersed as hot gases, as heat transferred to the parts of the engine, and partly as mechanical work that gets done in the engine to move the automobile. Finally, all the energy obtained by burning petrol gets dispersed in the surroundings as heat. The energy so dispersed is not accessible anymore to do any useful work and such energy is termed as the unavailable energy. Entropy is essentially the unavailable energy and the second law of thermodynamics states that the entropy of the universe always increases. This implies that every energy conversion adds to unavailable energy.

To do any useful work, we need concentrated sources of energy and the ability to control when and how much energy will be released and utilized. In this process, the concentrated source of energy gets dispersed and a part of the energy becomes unavailable. It is not possible to create a concentrated source of energy without giving a larger energy input that must come from some other concentrated source! For instance, Table 2.5: Energy Content of Fuels, mentions that hydrogen gas has an energy content of 123 MJ/kg. It has more stored energy per kg than petrol. However, hydrogen gas is not found in nature in the basic molecular form (H_2) required for combustion. To obtain the hydrogen gas by electrolysis of water, it is necessary to provide an energy input that is larger than the energy contained in hydrogen. Quite obviously, it is meaningless to carry out the electrolysis using electricity obtained by burning coal. There will be multiple energy conversions in the entire process and primary energy spent per kg of hydrogen so obtained will be much larger than the 123 MJ obtained by combustion of hydrogen. Petroleum is a concentrated source of energy that was created in the crust of the earth as a part of various geological processes. While it is possible to create petroleum products in the lab, the primary energy required to create such products will always be larger than what is obtained by their combustion.

These examples illustrate the fact that as energy sources are consumed, it is not possible to *create* them by processes that do not use some other concentrated source of energy. In other words, the depletion of energy sources is an irreversible process. The only energy source that won't deplete for about a billion years is the sun. Hence, the civilization cannot hope to be dependent on the fossil fuels forever. The decline of fossil fuels is inevitable and at some point in the future, it will be necessary to switch entirely to the use of solar energy, either in direct or indirect form.

Another important point that comes up based on the concept of entropy is that before proposing or advocating any new method of energy conversion or a new source of energy, detailed thermodynamic analysis is an absolute necessity. In absence of such an analysis, a wonderful idea in the lab may not provide any net energy.

To sum up, entropy and the second law of thermodynamics occupies a prime position among the laws of nature. Most of the struggle to find alternative sources of energy is essentially a struggle against entropy, which makes it so difficult. It is fairly common to read in newspapers about entrepreneurs, enthusiasts, and economists to have found a *solution* to the energy crisis. Yet, none of them turn out to be real solutions that are actually workable. It will be worthwhile to use an old quote from Sir

Arthur Stanley Eddington that sums up the problem that most enthusiasts do not realize:

"If someone points out to you that your pet theory of the universe is in disagreement with Maxwell's equations - then so much the worse for Maxwell's equations. If it is found to be contradicted by observation - well these experimentalists do bungle things sometimes. But if your theory is found to be against the second law of thermodynamics I can give you no hope; there is nothing for it but to collapse in deepest humiliation."

This quote may be modified in the current context to be:

"You may find the elusive Higgs Boson at CERN by putting together the most brilliant scientific minds and spending billions of dollars; but finding a solution to the energy crisis that somehow beats the second law thermodynamics is a hopeless quest."

The Concept of EROEI

EROEI is the Energy Returned On Energy Invested, during the lifetime of an energy source. This concept is best illustrated with an example:

Building a hydroelectric power station is a complex activity. Firstly, a detailed survey of the river basin is conducted to explore the feasibility of building a hydro-electric power station for the given geography and geology. On establishing the feasibility and completing the design work, it is necessary to start with building an access road for the construction work to start. A dam is built on the river and a reservoir is created. An electrical generation system is built to use the water from the reservoir to drive turbines to generate electricity. Finally, it is necessary to connect the generation system to an electric grid to distribute the electricity generated. Each of these activities require an energy input. This energy is spent *upfront*. The hydroelectric power station generates a certain amount of energy each year. The hydroelectric power station also has a certain expected lifespan. The ratio of the net energy generated during its lifespan and the energy spent upfront is the EROEI for the hydroelectric power station. In this case, net energy implies the energy generated during the lifespan minus any energy inputs during the lifespan as a part of operations, maintenance, and refurbishing.

EROEI is a very important concept not only for the renewable energy sources but also for the conventional ones like crude oil. An EROEI > 1 implies that there is some net energy available from the energy source. A value less than 1 implies that more energy is consumed in tapping the energy source than what it provides. For most energy sources, EROEI ranges from close to 1 to about 100.

Energy Up Front (EUF) and Energy Return Interval (ERI)

EUF and ERI are closely related to EROEI. EUF is the ratio of the amount of Energy spent upfront to manufacture and install an energy source, and the energy output during the first normal year of operation. For instance, if the amount of energy spent on the manufacture and installation of a solar photovoltaic energy system is 1000 MJ and the output from the system in the first year is 200 MJ, then the EUF is calculated to be:

$$EUF = \frac{1000}{200} = 5$$

The EUF is a measure of how energy intensive the production and installation of a given source of energy is. In general, a high EUF also implies a high capital cost for the energy source. The EUF also represents the Energy Return Interval (ERI). For the previous example, it will take 5 years to generate the energy that was spent in the manufacture and installation of the photovoltaic energy system.

EUF is a useful number for the energy sources that have very low operations and maintenance costs. Solar photovoltaic electricity, hydroelectricity, and wind energy are in this category. An energy source with a low value for EUF along with a long operational life with low maintenance will have a high EROEI. The EROEI for such a source can be calculated using the following formula:

$$EROEI = \frac{operational\ life\ in\ years}{EUF}$$

Among the conventional sources of energy, nuclear power tends to have a high EUF. Construction of a typical nuclear power station requires 5 to 10 years and involves significant energy use. Most oil and gas wells in the past had a low EUF, especially in the oil-rich regions of the world. The effort required to explore oil and to drill wells was small compared to the large output from the oil wells. Many of the old oil fields are large and have supplied oil for many years. The conventional oil and gas is perhaps the ideal source of energy from the perspective of EUF and EROEI! Newer oil and gas wells are, however, not so ideal. Many are in the sea and the oil is drilled from deep below the seabed. In some cases, the newer fields are declining rapidly and there is a doubt if they can supply the projected quantities of oil and gas. An example of this is the significant decline in output from the Krishna-Godavari basin offshore oil and gas field in India [ET 2012].

Traditionally, EUF has not been a number that is tracked by the energy companies. This was not a problem since most energy sources had a low EUF *and* a long operational life. However, most of the modern renewable sources of energy are not so ideal. They have a significantly large EUF. The EUF also translates to CO_2 emissions. It must be decided if a given source of energy actually saves any energy and CO_2! Secondly, when transitioning to a new energy source, high EUF becomes a major hurdle. For instance, for an energy source with an EUF of 7, it is necessary to spend 7 times the energy upfront as compared to the energy output in the first year. Where will this energy input come from? The transition may have been necessitated due to the expected dwindling of existing energy sources. In this situation, transition to a new source that requires a huge energy input from the existing energy sources is simply impossible or will require a very long time. During the long period of transition, there will be an immense scarcity of energy due to the additional demand generated by the transition which will certainly raise energy prices and hence the projects costs. Escalation of project costs may impact the economics of transition to the new source of energy. Hence, only an economic analysis can be very misleading when there is a possibility that the entire economy will be impacted by the energy transition and during the course of it.

Ideal Values for EUF and EROEI

An ideal source of energy would have a low EUF and a high EROEI. Let's say that the energy source has an EUF less than 1 and an EROEI greater than 50. Such a source will require a small energy input. It will return a large amount of net energy during its operational life and very likely make great economic sense as well. However, barring the fossil fuel deposits found in the first half of the 20th century, very few other energy sources are likely to be blessed with these numbers. The advocates of nuclear energy may object to this statement. However, that is a separate topic which will be covered in the chapter for nuclear energy. Now that such ideal sources of energy are no longer to be found what would be a reasonably good source of energy? This is a matter of opinion, but a source with EUF < 3 and EROEI > 7 can be considered a reasonably good one.

Using EUF and EROEI Instead of Economics

The normal practice in choosing an energy source for a given application is to use economics (return on investment) as the primary criteria. For instance, when someone decides to install solar panels on the rooftop, the decision depends on the following factors:

- Price of solar panels and installation costs.
- Interest rates (cost of money to be invested).
- Amount of energy delivered per year.
- Electricity tariff
- Expected life of the solar panel.
- Subsidies from government etc.

Such an analysis does not take into account the EUF or EROEI of solar panels and this approach makes sense for installations on a small scale. However, before advocating the large scale use of solar photovoltaic panels for generation of electricity, such an analysis is insufficient.

The conventional line of thought is to simply disregard the EROEI/EUF calculations and instead use economics as the means to estimate if solar PV electricity offers any net return on the investment made. There are some problems in this approach:

- The price of solar panels, associated electronics, batteries (if any), and the cost of installation are well known at the time of purchase. However, the price (or the tariff) for the electricity produced by the panel is a variable. In a grid connected system, the local electric utility company decides the tariff associated with a kWh of electricity generated and delivered to the grid. It is impossible to estimate how this tariff will change in the future. Life of a solar panel can be 25 years or more. Predicting the electric tariff over this time interval is not possible.
- Another variable is the inflation. The return earned by the energy produced by the panel needs to be adjusted for inflation. Predicting inflation over the next 25 years is mostly astrology and not economics!
- Interest rates will also vary during this interval, making it impossible to predict the real cost of capital.
- The price of the solar panel and the associated costs will change over the course of time, based on changes in the prices of various inputs. When considering transition to the use of solar panels as a major source of

energy, it must be considered if the production of solar panels can scale to meet demand on a very large scale that such a transition will produce. If solar panels do not produce any net energy, such a demand is likely to increase the energy costs which will escalate the input costs for the production.

All this makes it very difficult to determine if the solar panel actually provides any useful return on the capital invested. At the same time, this methodology does not provide a clue about the energy returns from a solar panel. The better option may be to make calculations on the basis of energy input and output. This implies the use of EUF and EROEI, to calculate which, it is necessary to determine the values for two attributes for the energy generated by solar panels:

- The energy used in manufacturing and installation of solar panels and the associated equipment. This is the energy input.
- The energy output per year of solar panels and the expected life of such a system

It is these parameters which will dictate if the Solar PV electricity is a good source of energy to consider transitioning to. Ideally, EUF < 3 and EROEI > 7 will make it a very attractive solution. Such an analysis will be helpful for any source of energy for which there is any plan of large-scale use.

The Problem of Calculating Energy Input

Estimation of energy output in a single year, or over the lifetime, is not very difficult for most sources of energy. The difficult part in the EUF and EROEI calculation is the estimation of energy input for the manufacture and installation of the energy source. I propose to use the energy intensity of economy for calculating the energy input.

Estimating the energy input for the manufacture and installation of an energy source is not an easy task. It may be possible to measure the exact amount of electricity or fuel that was directly used in manufacturing, transportation, and installation. However, it is the indirect energy inputs that are difficult to measure. For instance, manufacturing a solar panel requires a factory, manufacturing equipment, raw material and staff that ranges from scientists and engineers to janitors! It also requires capital in order to provide finance for the business. Thus, there are factors of production for any business: Capital, Labor, Infrastructure, and Raw Material. Each of them correspond to a certain energy consumption, as a result of the economic activity that they represent. In other words, all economic activity which is directly or indirectly related to the solar panel business corresponds to an energy input.

Energy Intensity of Economy and EROEI Calculation

Energy intensity is the average amount of primary energy consumed per unit economic output. It is measured in million Joules (MJ) per dollar (or any other currency). Every aspect of the economic activity requires energy. To calculate the energy intensity of economy, the total primary energy consumption in a year is divided by the GDP for that year. For instance, in the year 2009, energy intensity of the Indian economy was 0.34 MJ/₹. In the same year, energy intensity of the US economy was 7.5 MJ/$. In other words, every US dollar of economic output in the year 2009 corresponded to a consumption of 7.5 MJ of primary energy; and every rupee of economic output of the Indian economy corresponded to a consumption of 0.34 MJ of primary energy. This relationship between the economic output and energy consumption makes it possible to use the energy intensity of economy to estimate the energy input. This methodology assumes that in an open market with fair competition, the price of a commodity represents all direct and indirect inputs that go into the making of that commodity. The inputs include all the factors of production. If the commodity is manufactured on a sufficiently large scale, it should be possible to estimate the energy input by using the price of the commodity and the energy intensity of the economy as guides. At a macro level and for production on a large scale, such a methodology would be useful to estimate the energy input.

However, the calculation of energy input for an energy source is not as simple as just multiplying the price by energy intensity. Manufacturing processes of some energy sources tend to be more energy intensive. For instance, the process of manufacturing a solar cell requires a large direct energy input. It is necessary to account for the direct energy inputs separately. While an Indian Rupee (₹) corresponds to 0.34 MJ based on energy intensity, a rupee spent on buying electricity corresponds to about 2.7 MJ (based on electricity tariff in 2009 in Pune, India) of primary energy, almost eight times what the energy intensity suggests. Clearly, some assumptions need to be made to account for the direct energy purchases. Assuming that the cost of direct energy purchases for manufacturing and installation of an energy source is limited to about 12% (i.e. $1/8^{th}$) of the price, the total energy input for the energy source can be approximated by the formula:

$$\text{energy input} = \text{price of energy source} * \text{energy intensity} \times 2$$

The *magic number* 2 is specific to the Indian economy and is based on the cost of energy in India. For the US economy, this number is about 3. The energy input calculated with the use of this formula is an approximation. Yet, it is a useful tool for calculating the EUF and EROEI for an energy source.

Criticism of the Use of EROEI

Energy input and energy output are the two parameters that are used for the EROEI calculation. It is clear from the previous section that the energy input can only be approximated, while the energy output can be calculated fairly accurately. Quite naturally, the calculated value for EROEI (and EUF) is only an estimate. It is not possible to arrive at an exact value that would generically apply to an energy source due to the complexity involved. However, even the economic analysis inherently has inaccuracies and approximations. Hence, while the EROEI calculation provides a different perspective, it should be used along with an economic analysis, and is not a substitute for the economic analysis.

3. ENERGY, ECONOMY, AND HUMAN DEVELOPMENT

Once I challenged some friends to find an economic activity that does not involve the use of energy in any form, except biomass and the direct use of solar energy. Very quickly we came to the conclusion that it is almost impossible to find such an activity. Only in the remote tribal regions that still do not have any electricity or access to fossil fuels, could there be some such activities. These activities will be limited to gathering honey from the forest, growing Ragi, etc. Apart from these and similar unremarkable activities, every economic activity is dependent on the use of energy in some form. In fact, it won't be wrong to argue that the standard of living and human development itself is directly related to the availability of energy. It is quite interesting to explore the relationship between energy use, human development, and national incomes. The United Nations database provides excellent statistics that makes it possible to explore these aspects. Table 3.1 shows the total primary energy (TPE) consumption by economies in various parts of the world along with the 2008 Gross National Income (GNI) in US dollars, adjusted for purchasing power parity (PPP). The TPE consumption data was obtained from two sources:

- BP Statistical Review of World Energy, June 2011 [BP 2011]
- Energy Information Administration, US Department of Energy [EIA]

The GNI data was obtained from the UN database on World Development Indicators, World Bank [UN Data]. The data in this table is sorted on the basis of GNI, in the ascending order. The set of countries was chosen to include developing nations as well as the developed nations. A special emphasis is on India, China, US, and the developed nations in Western Europe.

Country	GNI 2008 (PPP US$)	TPE Consumption (TJ)	Energy Intensity (MJ/$ PPP)
Romania	290,339,425,899	1,638,000	5.64
Brazil	1,932,907,750,355	9,870,000	5.11
France	2,134,442,820,013	10,836,000	5.08
UK	2,218,207,841,493	9,030,000	4.07
Germany	2,952,423,602,513	13,734,000	4.65
India	3,374,885,012,516	18,522,000	5.49
Japan	4,497,715,219,976	21,672,000	4.82
China	7,984,041,157,696	87,360,000	10.94
USA	14,282,672,470,983	97,440,000	6.82

Table 3.1: Economy and Energy Consumption

It can be observed that the energy use increases along with GNI. Indian economy is almost 11 times larger than the Romanian economy and energy use in India is almost 11 times that in Romania. The table also calculates energy intensity for each economy. It can be observed that the energy intensity of most countries is close to about 5 MJ/$. For some countries, the relationship between GNI and energy use is somewhat skewed. This is reflected in the high number for energy intensity. A detailed discussion on energy intensity is available later in this chapter.

Size of Indian Economy and Energy Usage

In this book, there is a lot of emphasis on India due to the high economic growth rate in the previous decade and the expectation of continued growth in the coming decades. Analysis of economic data as well as the data on energy use in India provides for some interesting analysis. Indian economy has enjoyed this high rate of growth since the economic liberalization in the 1990s. During this interval, there has been a significant increase in the energy usage as well. Table 3.2 shows the change in GDP from 1999 to 2009 along with the TPE consumption.

Year	Inflation Adjusted GDP (Billion 2005 ₹)	TPE Consumption Trillion Joules (TJ)	Energy Intensity (inflation adjusted) (MJ/₹)	GDP (Billion ₹) (current for each year)	Energy Intensity (MJ/ current-₹)
1999	25,438	11,760,000	0.46	19,520	0.60
2000	26,464	12,390,000	0.47	21,023	0.59
2001	27,844	12,474,000	0.45	22,790	0.55
2002	28,893	12,936,000	0.45	24,546	0.53
2003	31,312	13,272,000	0.42	27,546	0.48
2004	33,910	14,448,000	0.43	32,392	0.45
2005	37,064	15,204,000	0.41	37,065	0.41
2006	40,562	15,918,000	0.39	42,840	0.37
2007	44,469	17,262,000	0.39	49,479	0.35
2008	46,746	18,522,000	0.40	55,744	0.33
2009	50,324	19,698,000	0.39	62,312	0.32

Table 3.2: India: Economy and Energy Consumption

The second column of the table contains the data for GDP, inflation adjusted to the year 2005. It is interesting to note that the inflation adjusted GDP increased by 97.8% from 1999 to 2009, and during the same interval, energy use increased by 67.5%. The fifth column contains the data for GDP that is not adjusted for inflation. This GDP increased by 219% in 10 years. Clearly, this growth in GDP is an illusion as a significant portion of it is consumed by inflation. It is the inflation adjusted GDP that tracks the energy use.

Data Sources:
1. UN Statistics Division: National Account Estimates of Main Aggregates [UN Data]
2. BP Statistical Review of World Energy , June 2011 [BP 2011]
3. BP Statistical Review of World Energy , June 2010 [BP 2010]

Energy Intensity of Economy

Energy intensity of economy is a number that specifies the energy use per unit economic output. Energy use is measured in Million Joules (MJ) and the economic output is the GDP, measured in the national currency. The GDP can either be inflation adjusted to some previous year or it could be based on the current year. To compare the energy intensity of different economies in the world, energy intensity is measured in terms of MJ per US dollar of economic output (MJ/$). In this book, the US dollar is used as the common currency to express the GDP of every nation. Conversion from national currency to US dollars is based not only on the market exchange rate but the Purchasing Power Parity (PPP) factor is also applied to get a realistic comparison. The GDP is also adjusted for inflation to calculate the energy intensity. Such a methodology makes it possible to compare the changes in energy intensity over the years. Table 3.1 shows the energy intensity of various economies in the world in the year 2008. It can be observed that the energy intensity of many nations is around 5 MJ/$.

Energy intensity is an important concept. It describes how efficient (or wasteful!) the economy of a nation is. It also provides some idea about how developed a nation is as well as about the contributors to the economy. Energy intensity is dependent upon following factors:

- Constituents of economic output: Manufacturing processes for certain products are highly energy intensive. If a major part of economy consists of such products then such an economy is bound to have a high energy intensity. The Chinese economy has a high energy intensity since it is heavily dependent upon the manufacturing industry that makes it the manufacturing hub of the world.
- Efficiency of processes used for production: Use of modern technology and efficient processes lead to energy savings, reducing the energy intensity.
- State of the Economy: Economic growth with low inflation tends to increase the energy intensity as consumers tend to conserve less. However, during the periods of recession and high fuel prices, there is a tendency to conserve. It can be recalled that the gas-guzzling, sports utility vehicles were always in great demand in USA. Demand for these vehicles reduced in the recent recession of 2008 and even after the end of boom period for the dot-com companies. However, in India, where the economic boom is continuing, demand for such vehicles is on the rise from the general consumer.

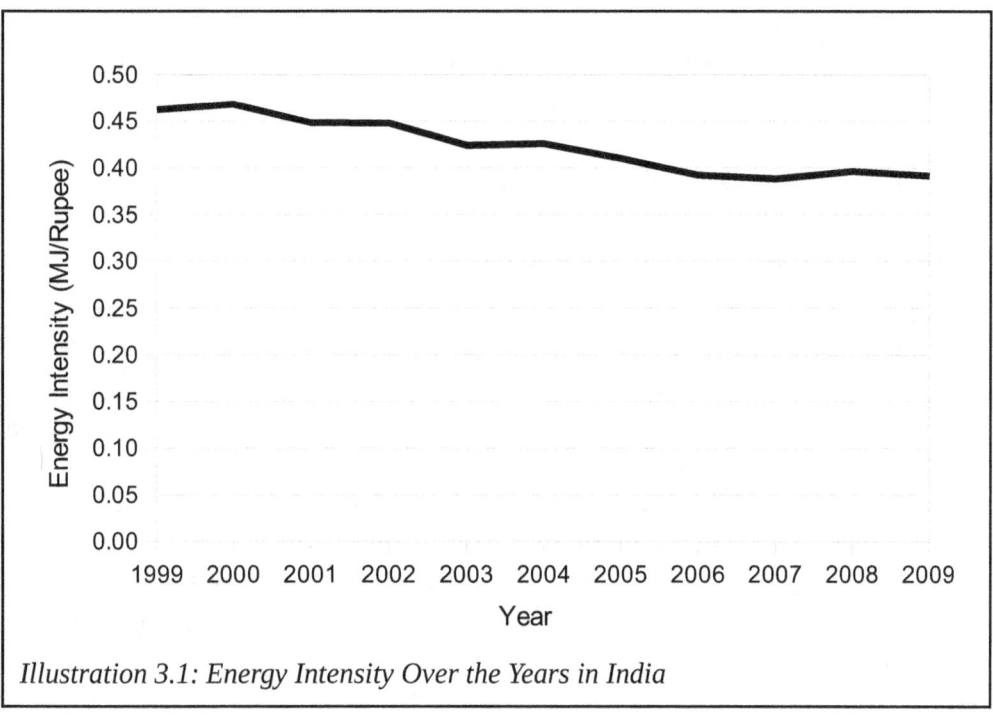

Illustration 3.1: Energy Intensity Over the Years in India

- Availability of Energy Resources: Availability of cheap energy leads to wastage. Saudi Arabia had an energy intensity of 12.9 MJ/$ in 2010 (inflation adjusted to 2008). Compare this with an energy intensity of 5.4 MJ/$ for India in the same year. In a country like India that has a huge population and scarce energy resources, there is a culture of conservation and reuse.
- Climatic conditions and geography: Extreme climatic conditions result in the use of energy for mostly unproductive uses such as air conditioning and water heating. A large country with sparse population will naturally require shipment of goods over long distances. The energy required for transportation over long distances is an unproductive use that leads to a higher energy intensity.
- Culture: American culture encourages super-sizing of consumption, whether it is single family homes, cars, or meal portions. It is no wonder that American economy has a higher energy intensity as compared to India, in spite of having access to the most modern technology as well as outsourcing a significant amount of industrial production to China. Japan has a much lower energy intensity, which at least in part, is a result of Japanese culture. A culture that

encourages, Reduce, Reuse, and Recycle (RRR) of materials tends to have a lower energy intensity. India is an example of such a culture. Traditional values have always encouraged the RRR policy. Most manufactured goods have a tremendous amount of latent energy content. The Reuse and Recycle policy makes certain that this energy is not wasted.

Energy Intensity: Change Over the Years

The data for the Indian economy, presented in Table 3.2, indicates that energy intensity has reduced slowly over the years. In 1999, the energy intensity of the Indian economy was 0.46 MJ/₹. By 2006, it reduced slightly to 0.39 MJ/₹. Illustration 3.1 shows the change in energy intensity over the years for the Indian economy. The change is likely due to a shift towards the service sector in the last decade and due to improvements in energy efficiency. Economies based on the service sector tend to be less energy intensive as compared to those dependent upon manufacturing and metallurgical processing. From 2006 to 2009, the energy intensity has not changed much, though the economy has grown by 24%. Thus, during this interval, the real growth in GDP is directly proportional to the increase in energy consumption. Hence, every incremental increase in economic output requires a corresponding increase in energy consumption. While this may seem to be a challenge, it is a relief that the energy intensity has not increased. Economic growth, reduction in poverty, and rising middle class are a perfect recipe for growth in the household consumption of energy. Such a rise in consumption normally leads to an increase in energy intensity. Yet, there is no evidence of such an increase in the case of India. Energy prices have certainly played a role in not allowing the energy intensity to increase. For instance, in most Indian cities, scooters and motorcycles are the preferred mode of personal transport, apart from the public transport that consists of suburban trains and buses. The price of petrol in India is so high, thanks to the extraordinarily high taxes, that it is not very cost effective to use a personal car as the daily mode of transport for most people. Scooters and motorcycles are, of course, very energy efficient as compared to cars. High energy prices lead to frugality and efficient use of energy resources.

Energy Intensity: Conclusions, Criticism, and Importance

There are some key conclusions to draw from the discussion on energy intensity:
1. For any real economic growth to occur, the available energy must increase.

2. Inflation is dependent upon the non-availability of energy for real economic growth.

3. Every dollar in circulation is a claim on energy, meaning that the demand for energy will rise as the amount of money in circulation increases. All the creation of money that takes place in a growing economy results in an increased demand for energy.

Though energy intensity is such an important tool, there are some problems in its effective use. One criticism about the calculation of energy intensity is the inherent possibility of some error in the data that is used for the calculation. For instance, the methodology of GDP calculation is certain to differ between different countries. Likewise, the exact measurement of energy consumption is difficult due to two factors:

a. Use of biomass as a source of energy is common in developing nations. Biomass mainly consists of firewood, crop residues, cow-dung etc. Measuring the exact amount of biomass used as a source of energy is obviously impossible.

b. Methodology for the calculation of primary energy for hydroelectricity and nuclear energy assumes a certain efficiency of thermal power stations. This efficiency is used to calculate the primary energy as if thermal power plants were used to generate the same amount of electricity. Such a calculation misrepresents the data for countries where hydroelectricity and nuclear energy are the main sources of electricity. For instance, hydroelectricity is the main source of energy in Norway and the use of coal is negligible. Similarly, France uses nuclear energy to generate most of the electricity. The methodology of calculation of primary energy unfairly penalizes Norway and France to report a higher value for energy intensity.

Another criticism of energy intensity is in its use as a tool to compare economic efficiencies of different nations. Energy intensity calculations do not factor in variables such as the size of the country, weather, and constituents of the economy. As a result, energy intensity cannot be used as a stand-alone tool for comparing different countries.

In spite of this criticism, energy intensity of economy is a very useful tool. For any large and complex country like India, it provides a wonderful macro mechanism to predict the future energy requirements of the growing economy. It also functions as a tool to calculate parameters such as, EROEI, and EUF. More importantly, measuring energy intensity over the years provides a good understanding about the development of a country as well as about the constituents of the economy. Every nation should strive to achieve a progressively lower energy intensity. Any economy that uses highly efficient technology for manufacturing processes and discourages waste at all levels will be able to achieve a lower energy intensity, despite continued economic growth.

It is also certain that a lower energy intensity will help reduce energy imports and hence enhance energy security of any nation that is dependent upon such imports. Lesser energy imports imply that the energy wastage that occurs in transportation is avoided. An added benefit of energy security may even be a reduction in the military requirements and hence the associated consumption of energy. For instance, the US Department of Defense (DOD) accounted for 93% of energy consumption by the US Government, and the oil consumption by DOD was higher than Sweden in 2005 [Lengyel 2007]. This is only the direct consumption of energy. The indirect consumption can be calculated by multiplying the defense spending (excluding direct energy purchases) by energy intensity. It is expected that the indirect consumption of energy will be much higher than the direct consumption.

Exponential Growth of Economy and Energy

"The greatest shortcoming of the human race is our inability to understand the exponential function." – Dr. Albert Bartlett

Every finance ministers pledges to follow economic policies that lead to a high growth rate while keeping inflation at bay. Indian economy has experienced rapid growth in the last fifteen years. In the recent years, the growth rate of Indian economy has been about 7 to 9%. An annual growth rate of 8% corresponds to a doubling time of about 9 years for the economy. Given the growth rate, we can mathematically figure out the GDP for any year in the future using the following formula:

$$GDP_n = GDP_0(1 + r)^n$$

In this formula, GDP_0 is the GDP for the base year. GDP_n is the GDP for the nth year from the base year, and r is the growth rate of the economy. In mathematics, this formula is an example of an exponential function. It is represented graphically for the Indian economy in Illustration 3.2.

If the Indian economy grows at an annual growth rate of 8%, GDP for India will rise from ₹ 50 trillion in 2009 to about 250 trillion in year 2030. The beauty of an exponential function is that it results in very rapid growth in a matter of few years. An yearly growth rate of 8% translates into 400% growth in 21 years! Such an economic growth indeed has very positive implications on the standard of living and poverty

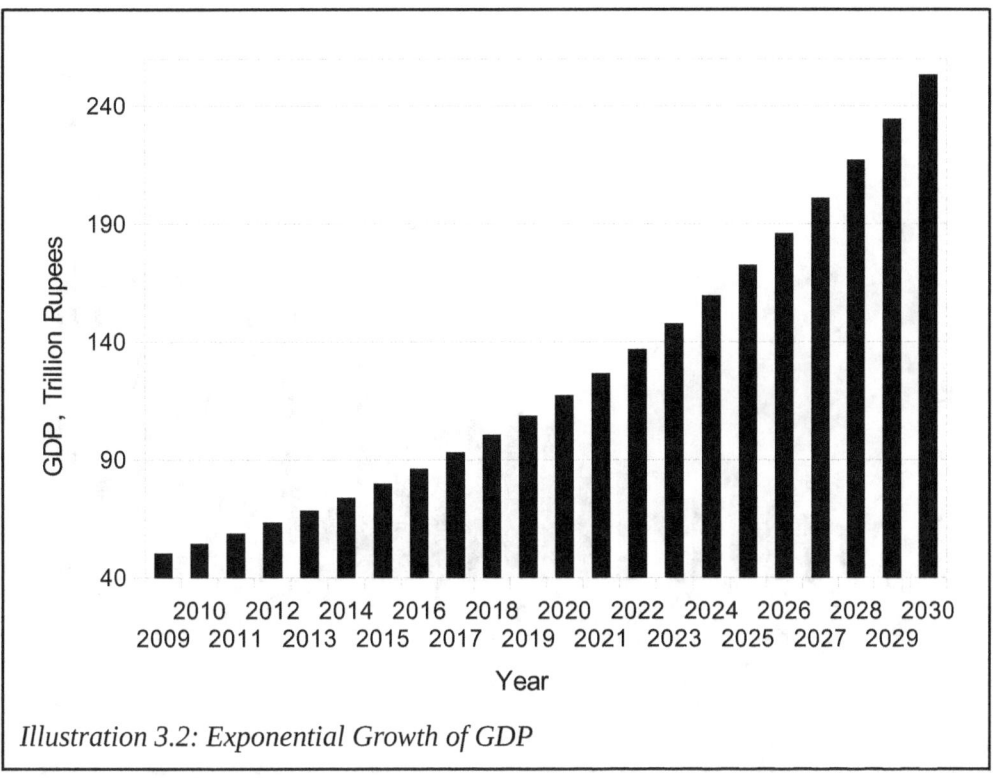

Illustration 3.2: Exponential Growth of GDP

reduction. It is no wonder that every economist and finance minister wants growth to follow such a trajectory. However, the question is if energy resources will be available to support such an economic growth. It was discussed earlier in this chapter that the energy intensity of economy does not change much over the years, and every dollar in circulation is a claim on energy. There is also a tendency for higher domestic consumption of energy due the affluence that economic growth brings in [Smil 2005]. Yet, in the case of India, there is little chance of an extreme increase in the domestic consumption due to energy scarcity and high energy prices. Thus, to sustain an 8% growth rate of economy, energy consumption must grow at a similar rate. Hence, the TPE consumption for the Indian economy can also be projected with the same formula:

$$TPE_n = TPE_0(1 + r)^n$$

Illustration 3.3 is a graphical representation of this formula, assuming that energy consumption grows at 8% per year.

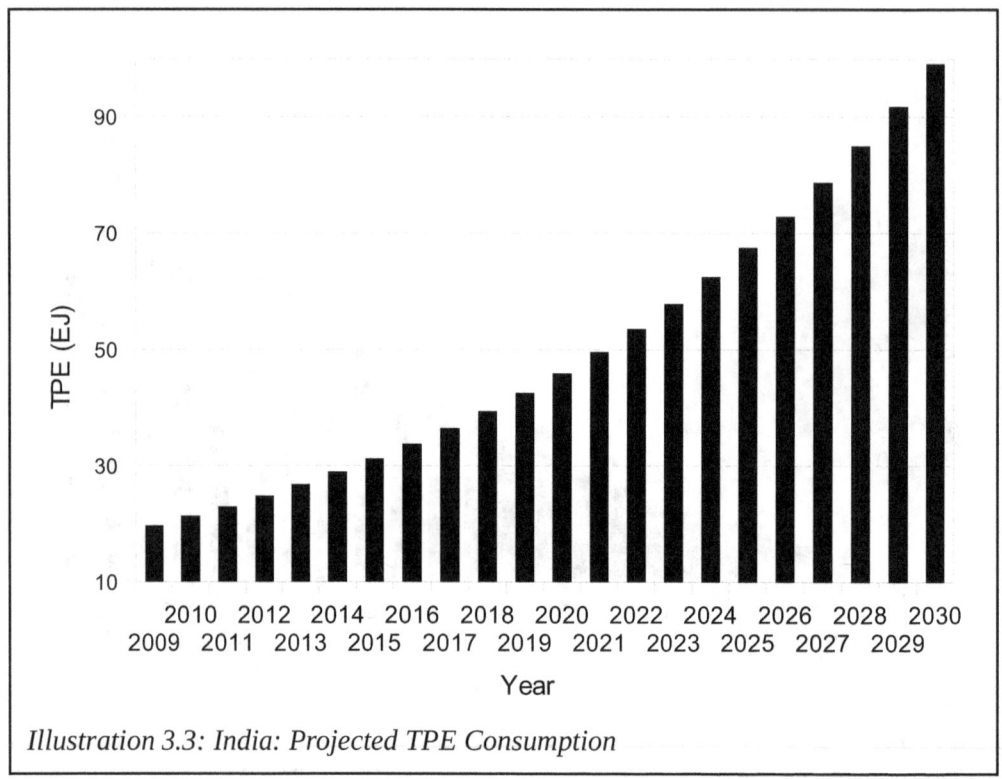

Illustration 3.3: India: Projected TPE Consumption

According to this figure, the TPE consumption in India will need to be about 100 EJ in 2030 to support the corresponding economic growth. A TPE of 100 EJ is more than the TPE consumption in USA in year 2009! The Chinese energy consumption is already approaching the energy consumption in USA and may well have overtaken it in 2010, according to the BP Statistical Review of World Energy, June 2011. Does the world have energy resources to support another USA? In other words, how much growth in energy consumption can be supported with the available resources? This is the question that will be explored further in this book. What if a conscious effort is made to reduce the energy intensity of economy so that the energy requirement grows at a rate lower than the economic growth? This is an incredibly difficult feat to achieve. Assuming that a 6% growth in energy consumption is sufficient to sustain an 8% economic growth, the energy requirement for India in the year 2030 is calculated to be 67 EJ. This is still 240% more than the consumption in the year 2009. While an exponential function is wonderful for economic growth, exponential rise in energy requirement is difficult to support on our finite planet.

Economic Growth in Developed Nations

Growth for the sake of growth is the ideology of a cancer cell --- Edward Abbey

Economic growth is a necessity not only in developing nations but also in the developed countries. The monetary system cannot function without economic growth. To understand this concept, it is first necessary to understand how money is created. Money is created every time when a bank gives out a loan. The loan could be to an individual, a company, or to the government itself. In other words, all money is loaned into existence. Imagine that there is a single bank in the city and a person opens a bank account with a $100 deposit. The bank may retain about 10% of this deposit as cash and loan out the remaining $90 to someone seeking a loan. With this loan some goods are purchased and the money transfers to a third person who deposits the $90 back into the bank in a separate account. The bank will again loan out $81 out of this deposit of $90. This process will continue until the bank has all of the $100, deposited originally, in its cash reserve, and has loaned out $900. In this process, the bank has practically created money using the original deposit. Money gets destroyed when the loans are paid back. In the previous example, when the loans worth $900 are paid back, there is a corresponding destruction of money. However, all loans carry an interest, and with a 10% interest rate, it will be necessary to payback $990 in a year. Thus, even after the loans are paid back, an extra $90 remains in the economy forever.

A couple of key conclusions can be drawn:

1. All money corresponds to some debt.
2. To pay the interest on the debt, it is necessary to generate more debt. As a result, debt must grow at least at a rate equal to the interest rate on the existing debt. This is an example of exponential growth and as a result, money supply and economy will need to grow exponentially.

It was stated earlier that all money is a claim on energy. An exponentially increasing money supply is an exponentially increasing claim on energy resources. Thus, it is not only the developing nations but even the developed nations are expected to stake an exponentially increasing claim on energy resources. In the last decade, however, developed nations have outsourced a lot of energy intensive and polluting manufacturing activity to China. This is one of the reasons that the energy intensities of the economies of the developed nations has reduced, and at the same time the energy intensity of Chinese economy is abnormally high. From the year 2000 to 2010, energy consumption in OECD countries increased from 5435 MTOE to 5568 MTOE, a very modest increase. Adding the energy consumption in China to these numbers, there was actually an increase in energy consumption during the same time period from 6473 MTOE to 8000 MTOE. It is also important to note that during the recession of 2008-09, energy consumption decreased in OECD countries from 5715 MTOE in 2007 to 5378 MTOE in 2009.

This data indicates that the monetary system will demand an exponential economic growth that will need an exponentially increasing energy supply. Even the developed nations cannot escape this fact. On a finite planet, there will be a limit to the energy supply, unless a new energy source or a new method of harnessing energy efficiently is discovered. Until then, a clash between the requirements of economic growth and the limits of energy supply is inevitable. It is very likely that the recession in 2008-09 was at least partly caused by such a clash.

Energy and Human Development

It is very interesting to analyze the linkage between energy consumption and human development. Unites Nations (UN) defines the Human Development Index (HDI), which is a measure of how developed a nation is. This index consists of three components:

1. Life expectancy at birth
2. Education
3. Gross national income

The progress of a nation is calculated using a formula based on data for these three factors. The formula results in a number between 0 and 1. A number close to 1 indicates a country with very high human development. In 2010, Norway ranked first with an HDI of 0.938. USA ranked 4th with an HDI of 0.902, and India was placed 119th with an HDI of 0.519. At the bottom were countries from Sub Saharan Africa. For instance, Niger was ranked 167th with an HDI of 0.261. There is a correlation between HDI and the per capita energy consumption, electricity production, and the Gross National Income (GNI). Illustration 3.4 is a plot of HDI vs per-capita GNI, measured in US$ for the year 2010, with an adjustment for purchasing power parity (PPP). The GNI is inflation adjusted to the year 2008. It shows that at very low levels of GNI, HDI is very low. As GNI increases, HDI rises rapidly. A *reasonable* HDI of 0.7 requires a GNI of about $10,000. It can be observed that the relationship between GNI and HDI is logarithmic. Hence, reaching an HDI of 0.8 from 0.7 requires doubling of GNI to about $20,000 and an HDI of 0.9 requires a GNI of $40,000. Thus, at high levels of GNI, the rise in HDI is less pronounced. There are some countries,

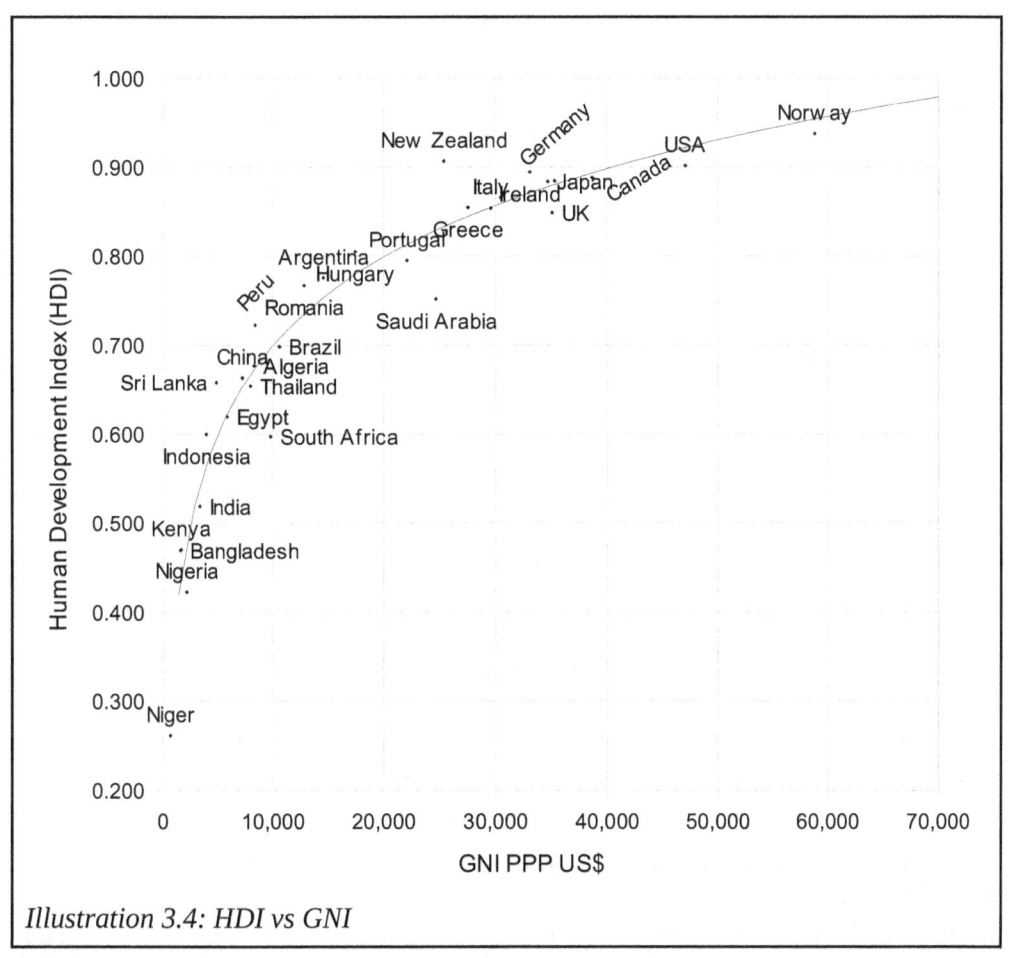

Illustration 3.4: HDI vs GNI

Sri Lanka and Peru for instance, that achieve a higher HDI with a relatively low GNI. These countries are located above the curve in the Illustration 3.4. At the same time, South Africa and Saudi Arabia make a poor use of their national income towards achieving a high HDI and are located below the curve. India has a long way to go and is still at the lower end of the curve. At this point, an increase in GNI for India should lead to a rapid rise in HDI, provided that the fruits of rising national income reach the masses in an efficient way. Indian economy has been growing at a rate of 7 to 9% per year for the past few years. Such a growth rate will improve the GNI to more than $10,000 in about 13 years, making possible the achievement of a reasonable HDI of 0.7.

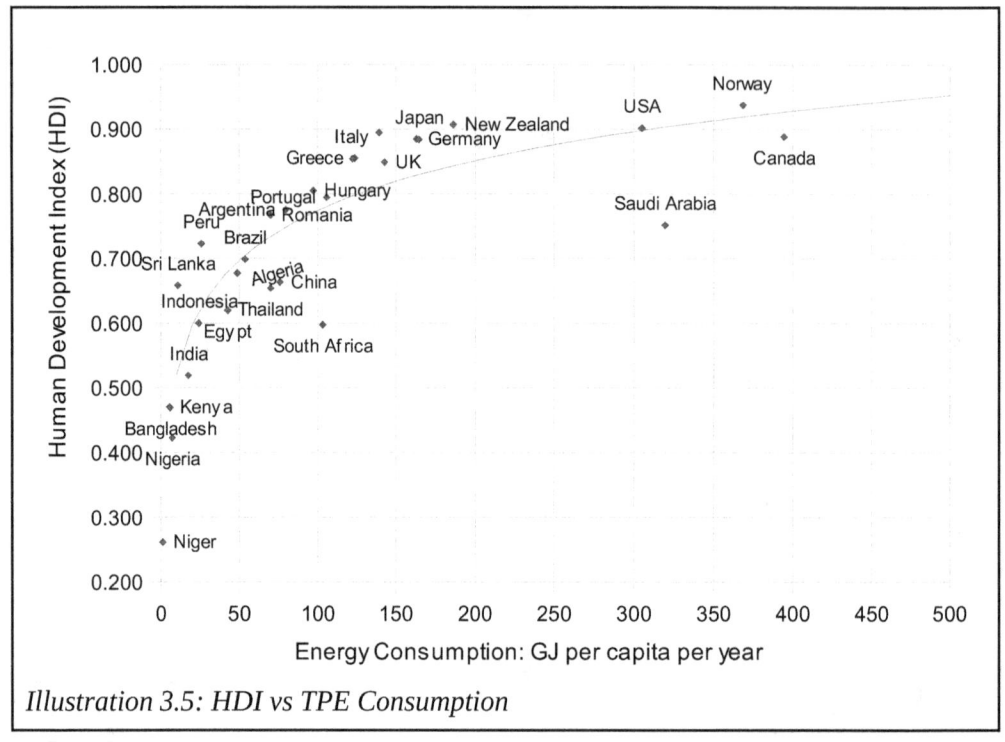

Illustration 3.5: HDI vs TPE Consumption

For the purpose of this book, it is even more important to analyze the correlation between energy consumption and HDI. Illustration 3.5 shows the plot of HDI vs TPE consumption, measured in GJ per capita per year. Once again, the logarithmic relationship between HDI and TPE consumption is quite evident. Until about 50 GJ per capita, HDI rises very quickly with energy consumption. After that, the rise in HDI slows down, and after about 200 GJ per capita, HDI does not rise much with energy consumption. There are countries like Sri Lanka and Peru, which use energy quite efficiently towards the achievement of a reasonable HDI. At the same time, South Africa and Saudi Arabia do not achieve a very good HDI, despite consuming a very large amount of energy. Yet, for most countries, the HDI-TPE correlation is very obvious. Consuming energy beyond 200 GJ per capita is mostly wasteful as it results in almost no improvement in human development. It may be argued that the extremely cold weather conditions in Canada, Norway, and some parts of USA result in a very high consumption of energy. However, there are examples to the contrary such as, UK, Japan, and Germany, who use energy much more efficiently towards the achievement of a high HDI. It can be concluded from the data presented in Illustration 3.5 that to achieve a *reasonable* HDI of 0.7, about 50 GJ of energy consumption is necessary per

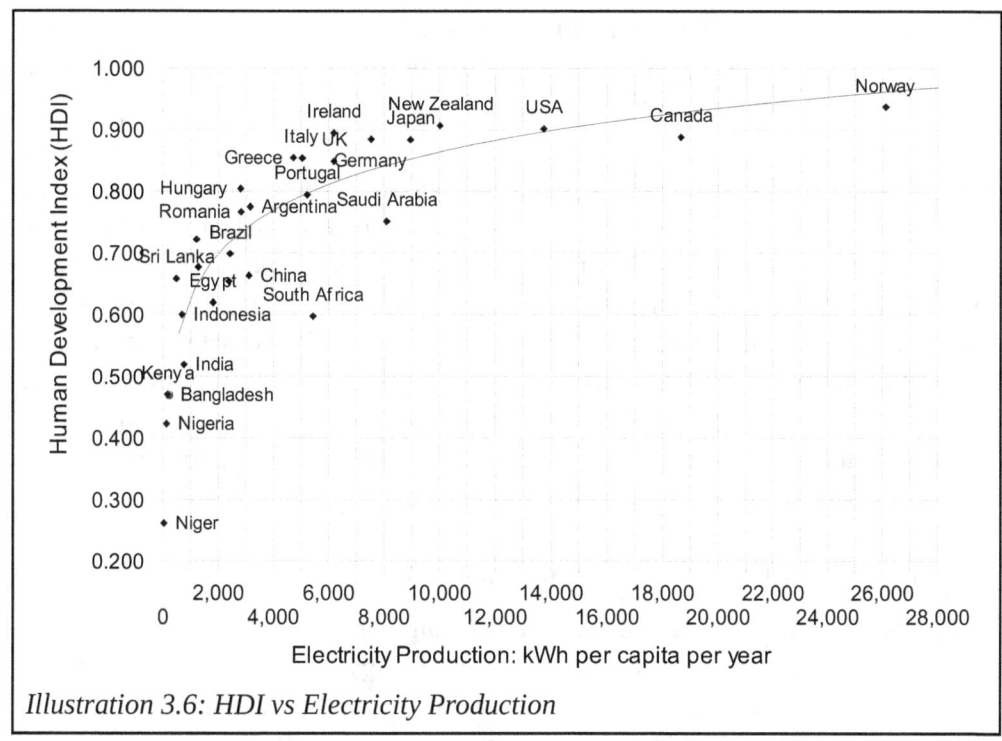

Illustration 3.6: HDI vs Electricity Production

capita, per year. A more respectable HDI of 0.8 will require about 100 GJ per capita, per year.

Another important correlation to analyze is between the HDI and electricity production. Electricity is a modern source of energy and is closely related to development. Illustration 3.6 shows the relationship between HDI and electricity production, measured in kWh per capita per year. Once again, countries with very low production of electricity have a very low HDI. Up to a production of about 1500 kWh per capita per year, the HDI rises quickly along with the rising production of electricity. After that, the rise in HDI is slower. After about 6000 kWh per capita per year, and an HDI close to 0.9, there is no rise in the HDI with a rising electricity production. To achieve a *reasonable* HDI of 0.7, about 2000 kWh per capita per year should be sufficient. An HDI of 0.8 should be achievable with an electricity production of 3000 kWh per capita per year. Some relevant examples are, Hungary (HDI 0.81, electricity 2817 kWh/capita/year), Romania (HDI 0.77, electricity 2837 kWh/capita/year), and Argentina (HDI 0.78, electricity 3167 kWh/capita/year). India has an HDI of 0.52 with an electricity production of 756 kWh per capita per year. There is no suggestion here that simply raising the production of electricity or consuming more energy will raise HDI. However, in order for the HDI to rise, a number of changes need to take place in the infrastructure of a country. It is this infrastructure that requires a corresponding energy consumption.

One criticism of the use of HDI is that it is an inadequate measure of human development. The UN considers only 3 factors: income, literacy, and life expectancy for the HDI calculation. It does not include important factors such as, democracy, freedom, distribution of income, crime rate, morality, and happiness. In the existing system, a dictatorial regime may score high on HDI, provided its citizen live a long life that may be devoid of freedom. Despite this criticism, HDI is useful when used as a relative measure. For instance, India may want to improve on HDI from 0.52 to 0.7 in the next 15 years. Using the correlation of HDI with energy consumption and GNI, it is possible to predict the economic growth rate and energy consumption that will be necessary to achieve the improvement in HDI.

Estimating Energy Consumption in India in 2030

Energy consumption by India and China are certain to impact the global energy prices as well as the emissions of greenhouse gases. Energy consumption in China has increased rapidly over the past two decades and it is now the top consumer of energy in the world. In the case of India, economic reforms and GDP growth over the past decade have resulted in a significantly increased energy consumption. Yearly growth rate of energy consumption in China is around 8.9% over the past decade and in India it is about 5.9%. China consumes about 20% and India consumes about 4% of the total energy consumed in the world. In this section, consumption need for India will be analyzed as that is where the next major growth in energy consumption is likely to occur. This growth will not be for the sake of growth but improving quality of life, as reflected in the HDI, will be the motivation.

In 2010, India had an HDI of 0.52 with a per capita energy consumption of 18 GJ/year and an electricity production of 756 kWh per capita, per year, with a GNI of $3,337 (PPP US$ 2008) per capita. A reasonable HDI target for year 2030 is 0.7, which is classified by the UN as *High Human Development*. It is necessary to estimate the GNI, TPE consumption, and electricity production to achieve this HDI number. It was discussed earlier that a per capita GNI of $10,000 is necessary to achieve an HDI of 0.7. In the year 2030, population of India is likely to reach 1.5 billion, which is the medium variant estimate from the United Nations Population Division. This population, with a per capita GNI of $10,000 (PPP, inflation adjusted to 2008), implies a total GNI for India of 15 trillion dollars. In 2008, GNI for India was 3.37 trillion dollars. Raising this to 15 trillion dollars over a course of 22 years requires an annual growth rate of about 7%. Remember, this is the actual growth, after subtracting the inflationary part of the growth. Considering the experience in the 2000 – 2010 decade, such a growth rate is certainly possible, as far as economics is concerned. The real question is about the energy consumption required by such an economic growth. Assuming an energy intensity of 5 MJ/$, the GNI of 15 trillion dollars will require an energy consumption of 75 billion GJ, which is the same as 75 Exa Joules (EJ)!

Another method to estimate the energy requirement is based on the correlation between HDI and energy consumption. Analysis of HDI data indicates that an energy consumption of 50 GJ per capita is necessary to achieve an HDI of 0.7. Hence, in the year 2030, a population of 1.5 billion will require an energy consumption of 75 EJ, which tallies with the calculation based on the HDI-GNI correlation. Such an increase in consumption requires an annual compounded growth rate of 6.5% for energy consumption. An energy consumption of 50 GJ per capita for a GNI of $10,000 per capita corresponds to an energy intensity of 5 MJ/$. Such an energy intensity corresponds to a reasonably developed economy as would be expected with an HDI of 0.7. The real problem is in finding the energy resources that will allow a consumption of 75 EJ. It is interesting to note that energy consumption in USA stood at about 97 EJ in 2008. Harnessing the conventional energy resources to increase energy consumption in India to 75 EJ is a significant challenge.

Another important aspect is to estimate the electricity production requirement in the year 2030. An electricity production of 2000 kWh per capita per year is observed to be necessary to achieve an HDI of 0.7. As a result, in the year 2030, for a population of 1.5 billion, it will be necessary to generate 3000 billion kWh of electricity. Assuming an average capacity factor (aka PLF) of 0.7 for power plants, it can be calculated that a generation capacity of 490 GW is necessary. This is a phenomenal increase from an installed capacity of 175 GW in 2008. In the following chapters, different energy sources will be analyzed for their potential to meet the projected energy demand.

Before moving forward, let's calculate the energy consumption that will be necessary in India to attain an HDI of 0.8. It was discussed earlier that an energy consumption of 100 GJ per capita is observed to be necessary to achieve an HDI of 0.8. In the year 2030, for a population of 1.5 billion, India will need to find energy resources to consume 150 EJ of energy per year. This is 50% more than the consumption in USA in the year 2010! A similar calculation for electricity suggests a capacity requirement of 735 GW for electric power in the year 2030!

With this data, it should be fairly obvious that a reasonable target for energy consumption should be based on achieving an HDI 0f 0.7. Achieving an HDI of 0.8 is nearly impossible for a population of 1.5 billion, using the conventional model of development. In summary, the energy requirement projection for India in the year 2030 to achieve an HDI of 0.7 is:

> **India: Energy Requirement in the Year 2030**
> - TPE Consumption: 75 EJ.
> - Electric Power Plant Capacity: 490 GW

Argument Against Using Statistics

Calculations in the previous section are certainly demoralizing. Meeting the projected energy requirement is quite difficult which implies that a huge population will never see the light of development and will likely languish in poverty. However, it must be kept in mind that such dire predictions are purely based on statistics and the past experience. Statistics cannot be used as a predictor of future when technological breakthroughs change the way we live. Likewise, dramatic social changes that allow the poorest person to access the fruits of development can make a positive impact on the HDI numbers, while consuming low energy. Yet, it must be stressed that statistics cannot be ignored. Without a dramatic change in technology or social structure, past statistics will predict the future!

Current Use of Energy Sources

Now that the projection for energy requirement in the future is available, it will be useful to take a look at the energy sources that constitute the energy consumption in India and in the OECD countries. Illustration 3.8 is a pie chart that shows the energy consumption in India in the year 2010 by each resource type. The three fossil fuels, viz. coal, oil, and natural gas, account for 94% of the total energy consumption. Hydroelectricity and other renewable sources of energy account for just 6% of the energy consumption. Note that the percentages in the chart add up to 101% due to rounding errors. It will be possible to extrapolate this data along with the projection for total energy consumption in the year 2030 to predict the requirement of fossil fuels in that year. Illustration 3.7 shows that 83% of the energy that the OECD countries consume is derived from fossil fuels. Hydroelectricity and other renewable resources

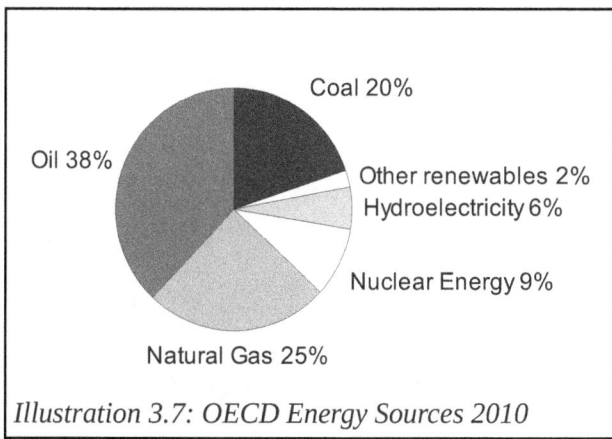

Illustration 3.7: OECD Energy Sources 2010

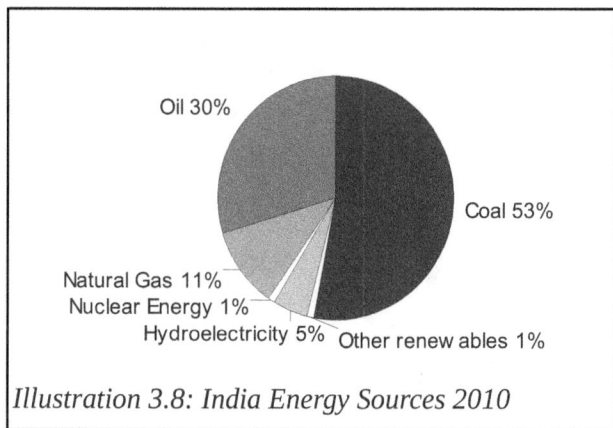

Illustration 3.8: India Energy Sources 2010

account for just 8% of the energy consumption. This is not very different from India. The primary difference between the scenario in India and in the developed world is in the use of coal and natural gas. The developed world prefers the use of natural gas for electricity generation, while in India, natural gas resources are scarce and less explored, while coal is available in relative abundance. Secondly, in spite of all the promise, dependence on nuclear energy in OECD countries is very low, at 9%. India is embarking upon an ambitious nuclear energy program to increase the production of nuclear energy. Currently, nuclear energy accounts for only 1% of the energy usage in India. It will be important to analyze if nuclear energy can meet the projected demand in the future.

Before closing this chapter, it will be interesting to note some data from China. The percentage of TPE consumption derived from coal increased from 61% to 70% in China from the year 2000 to 2010. In other words, a great part of the economic boom in China was based upon coal as the main source of energy. Is this sustainable in a world that is becoming increasingly aware about the consequences of global warming? Would replication of this Chinese model of economic growth be feasible for India or any other developing country? In the following chapters, the potential of various energy sources will be analyzed to estimate their capacity to meet the projected future demand for energy.

4. A FUTURE BASED ON FOSSIL FUELS?

The last 100 hundred years can certainly be classified as a century of fossil fuels. In this era, fossil fuels were the primary source of energy which brought about an unprecedented prosperity. An average middle class household now consumes the amount energy at the push of a button, that was available to only the super-rich, just about 150 years ago. Such a power would not have been possible without the use of fossil fuels. Yet, a majority of the world population still does not have access to this prosperity. There is an expectation that economic growth will lift millions out of poverty, and as discussed in the previous chapter, energy consumption will correspondingly increase. In this chapter, let us take a look at the fossil fuel sources, and if they can support the projected growth in energy consumption in the next 20 years. There will be a focus on India since that is the place where a major part of the growth is expected to take place. There will also be a focus on USA and China due to their status as the largest consumers of fossil fuels.

In the previous chapter, it was observed that 94% of the energy consumed in India was derived from fossil fuels. Even in the OECD countries, 84% of the energy consumed was derived from fossil fuels. Over the next 20 years, energy consumption in India is projected to rise to 75 billion GJ per year, which is almost 270% more than the consumption in year 2010. Using these numbers, it is possible to calculate the demand for fossil fuels from India. Before going into those details, it will be worthwhile to take a look at the energy consumption in China, since it is suggested many times that India follow China's footsteps. Over the last two decades, it was in China that a phenomenal growth in energy consumption took place along with economic growth.

China and Fossil Fuel Consumption

The Chinese economy achieved a phenomenal growth in the past two decades, registering almost 170% growth from the year 2000 to 2010 [UN Data]. During the same period, energy consumption in China increased by about 200%. This data implies that the energy intensity of the Chinese economy increased, probably due to the heavy dependence of economy on manufacturing and metallurgical processes and products. During this period, the dependence on coal actually increased, while in most developed countries the dependence on coal was falling. In 2010, China derived 70% of the Total Primary Energy from coal. Compare this with just 20% for the OECD countries and 53% for India. From the year 2000 to 2010, consumption of coal in China increased 3 times and the production of electricity also increased 3 times [BP 2011]. The consumption of natural gas was just 4% of the TPE consumption in 2010. The consumption of petroleum was 18% of the TPE consumption, a reduction from 29% in the year 2000, though there was an increase in consumption in absolute terms. The Chinese example clearly indicates that the economic growth in China was powered primarily by coal, and not by the more environment-friendly natural gas. As the Indian economy seeks to grow in the next two decades, there is a distinct possibility that it may also depend on coal, following the Chinese model.

Illustration 4.1: Traffic in Beijing, China.
Credits: Testing/Shutterstock.com

India: Projections for the Year 2030

In the previous chapter, projections were made for energy consumption in the year 2030 for India. The projections were based upon the goals to attain an HDI of 0.7 or 0.8. These goals require an energy consumption of 75 EJ and 150 EJ per year, respectively. It is possible to estimate the fossil fuel requirements based on these projections and the existing pattern of consumption. Table 4.1 projects the consumption of fossil fuels based on the current consumption pattern. The data for the year 2010 is presented in the second column of this table. The subsequent columns in this table are for the three projections for energy consumption in the year 2030: 75 EJ, 100 EJ, and 150 EJ. It is assumed that crude oil, coal, and natural gas satisfy the same percentage of energy requirement, as in the year 2010. The consumption numbers are specified in oil equivalent (MTOE) terms as well as using standard units.

	Current (2010) 22 EJ	Projected (2030) 75 EJ	Projected (2030) 100 EJ	Projected (2030) 150 EJ
Oil (million tonnes)	156	531	707	1,061
Oil (million barrels)	1,143	3,892	5,182	7,777
Natural Gas (MTOE)	56	190	254	381
Natural Gas (billion m³)	62	211	282	423
Coal (MTOE)	278	946	1,261	1,891
Coal (million tonnes)	734	2,497	3,329	4,992

Table 4.1: Projections for Fossil Fuel Consumption in India

Calculations in this table are based on the following conversion factors:
- 1 tonne of crude oil = 7.33 barrels
- 1 MTOE = 1.11 billion cubic meters of natural gas
- 1 MTOE = 2.64 million tonnes of coal. This number actually varies between 1.5 to 3, depending on the quality of coal. The approximation in this book is based on the data for India for the year 2010 [BP 2011].

These projections can be used as a basis to predict the demand for fossil fuels from India. It is possible to make similar projections for China and other developing nations. Even for the OECD countries, the expected rate of economic growth can be used as the means to project the consumption numbers. While the demand for fossil fuels from developing nations is increasing rapidly, as discussed in the previous chapter, even the developed nations need an increased consumption to keep their economic engine running. The question that needs to be addressed is, if the ever-increasing demand for fossil fuels can be met by the existing resources.

In India, supporting even the present requirement of consumption is a significant problem. The proven reserves of petroleum in India were at 1.2 billion tonnes in the year 2010 [BP 2011]. Even if these reserves are not depleted by the year 2030, the lower-end projection for consumption of 531 million tonnes will consume the reserves in less than 3 years! Unless there is an order of magnitude to increase in the reserves, the situation will be quite critical. A similar argument can be made for natural gas as well, for which the R/P ratio (see Glossary) stands at 28.5 years in the year 2010 [BP 2011]. The projected consumption of natural gas in the year 2030 will deplete the reserves in less than a decade. This data implies that India will continue to depend on imports of crude oil and natural gas to meet most of the domestic demand. The situation in the case of coal is certainly more comfortable than oil and natural gas, with an R/P ratio of 106 years in the year 2010 [BP 2011]. This data and the Chinese example suggests that there may be a very heavy reliance on coal in the future, as a driver of economic growth in India.

Coal, Oil, and Natural Gas: A Brief History of Transitions

Coal, Crude Oil, and Natural Gas are the three primary fossil fuels used by mankind today. While coal has been used for more than a millennium, it was only after the industrial revolution that the production and use of coal increased. In the 19th century, coal was the primary fuel that powered the newly industrializing society. Prior to that, all types of biomass, ranging from firewood to cow dung, was used as the primary source of energy by civilizations around the world. During the 19th century, discoveries of crude oil and natural gas wells were made. However, their use was not so widespread. Natural gas was used for municipal lighting in towns close to the source of natural gas. Petroleum, or Rock-oil, was used for heating homes and in cooking stoves. It was only after the invention of automobiles and the internal combustion engine, that the use of petroleum products increased by orders of magnitudes. Invention of electric bulb obsoleted the application of natural gas for lighting. At that point, the use of natural gas shifted to heating water and for heating homes. This required the pipelines of natural gas to be built, which was possible with improvements in pipe-making technology that allowed carrying gas at high pressure over long distances. This history shows three transitions of energy sources, as the primary drivers of the civilization:

1. The first transition was from biomass to the use of coal. This transition took place in the middle of the 19th century. Coal continued to be the primary source of energy for almost a century. This transition was mainly a result of the industrial revolution that used coal as the source of energy for machines.

2. The second transition was from coal to petroleum. This transition happened around the middle of the 20th century.

3. The third transition is still in progress, which is the use of natural gas as a replacement of coal for electricity generation and in heating applications. To a certain extent, natural gas is also replacing petroleum products in the use for transportation.

As things stand today, it is quite uncertain if the third transition will actually be able to replace/reduce the use of coal. During the last 10 years, from 2000 to 2010, the use of coal in China has increased more than 3 times. Coal supplies about 70% of the Chinese TPE consumption in 2010, while natural gas supplies only about 4%. This fact has reversed the trend (reduction in the use of coal), and in 2010, coal supplies 30% of the energy consumed in the world, as compared to 24% in the year 2000. During this time, the consumption of natural gas has remained steady at 24%. Petroleum has remained the primary source of energy, but its share has fallen from 39% in 2000, to 34% in 2010. Another interesting fact is that over the last couple of years, significant increase was achieved in the production of natural gas in USA with the shale gas fields ramping up production. This will certainly help reduce the use of coal for the production of electricity in USA. The next 10 years promise to be very interesting. Whether coal once again becomes the primary source of energy, or if the increase in the production of natural gas promotes it to be the number one source, is very difficult to predict.

Fossil Fuels: Characteristics and Use

Applications of fossil fuels are not restricted to the field of energy. They range from metallurgy to pesticides, and asphalt to fertilizers. In this section, characteristics of fossil fuels that result in specific uses in the field of energy are analyzed.

1. **Coal**: An important characteristic of coal is its relative abundance all over the world. This not only makes coal relatively inexpensive, but also a democratic source of energy. It is difficult for a small number of countries to control the

The Iran-Pakistan-India Natural Gas Pipeline

A pipeline from natural gas rich Iran to energy deficient Pakistan and India seemed like an ideal solution that would address the energy woes in South Asia. Such a pipeline would open the huge Indian market for Iran, while India can avoid the expensive imports and the infrastructure for Liquefied Natural Gas (LNG). Even Pakistan would benefit immensely from such a pipeline. In India, such a pipeline would allow the use of natural gas for generation of electricity as well as for urban transportation. Both these sectors are vital for the growing Indian economy. An added benefit would be the reduction in pollution and CO_2 emissions. Yet, the proposal for this pipeline is as good as shelved. Firstly, after the terrorist attacks on Mumbai on 26/11, no one in India would trust Pakistan to have the controlling tap for a critical, energy-supply line. Pakistan is a country in turmoil with militancy out of control. Passing an energy supply lifeline through Pakistan in too risky for India. Another aspect has been the US opposition to such a pipeline. Revenue generated by selling gas to India and Pakistan would be helpful for Iran to fight any embargo imposed by the US or UN for the Iranian nuclear program. A nuclear Iran is a geopolitical risk that US is not willing to take. As a result, there has been pressure from US on India and Pakistan to shelve the pipeline program. After signing the nuclear deal with US, India has as good as abandoned plans for such a pipeline. However, there is a new proposal now for piped natural gas from Turkmenistan to India, via Afghanistan, and Pakistan. The acronym for this pipeline project is TAPI and is reported to have blessing from the Americans [TOI 2012].

international trade in coal. In the case of India, the quality of hard coal is quite poor due to its high ash content and consequently a lower energy content. Yet, India produced almost 570 million tonnes of coal in 2010 and ranked only behind China and USA. Due to the solid nature of coal and a lesser energy density compared to crude oil, transportation of coal is expensive. As a result of these characteristics, the primary use of coal is the generation of electricity in thermal power stations. Such power stations are generally located close to coal mines. According to the International Energy Agency (IEA) statistics for world for the year 2009, about 41% of electricity was generated using coal. One problem with coal is the pollution that it causes. Also, a solid fuel is generally a bad choice for transportation applications as well as for applications that require wide distribution. As a result, the use of coal for domestic heating and cooking, and transportation has reduced significantly over the past 100 years.

2. **Petroleum** (crude oil) is the primary source of energy, supplying 34% of the TPE consumption in the world. The main advantage of petroleum and its distillates, such as petrol, diesel, kerosene, is the high energy density and the liquid form. This makes it the perfect fuel for transporting across continents as well as a perfect fuel for use in the transportation sector. All aircraft, almost all automobiles, and large ships, use petroleum products as the fuel. It won't be incorrect to say that local and global trade will come to a standstill if petroleum becomes unavailable. The problem with petroleum is the concentration of petroleum reserves in a few countries in small geographic regions. Saudi Arabia, Iran, Iraq, Kuwait, UAE, Qatar, and Venezuela control about 69% of the global reserves of crude oil. The most populous nations; viz. China and India, have limited reserves and while USA has relatively better reserves, extreme consumption requires USA to be dependent on petroleum imports. Most countries in Europe, while having high per capita consumption of petroleum, have low or dwindling reserves. It won't be wrong to say that most of the petroleum consumed in the world is imported! It is well known that obtaining control of petroleum reserves was one of the objectives behind several wars in the last 150 years. Lately, the price of crude oil has remained very high, above $100 per barrel. This makes it a very expensive fuel and rather unsuitable for the generation of electricity. Yet, no other fuel can match petroleum for its use in the transportation sector.

3. **Natural Gas** is the cleanest burning fossil fuel. Not only is the emission of pollutants, such as sulfur and mercury, the lowest for natural gas, but CO_2 emissions are also the lowest per calorie of heat. It is cheaper than petroleum though more expensive compared to coal in most cases. Its gaseous nature at normal temperature and pressure implies that energy content per liter is very low though energy content per kg is actually higher than petrol. This characteristic makes it a difficult fuel to transport and store. Where a pipeline is not available, transport of natural gas requires special cryogenic tankers or vessels that maintain natural gas in a liquid form. It is necessary to cool natural gas to about -160° Celsius to convert it into a liquid form. The vessel or tanker, and the storage facilities must maintain the gas at this temperature. Even the natural gas pipelines must be built to transport the gas at high

pressure. Natural gas can also be transported and stored in a compressed form, known as CNG. The containers need to be designed to sustain high pressure (more than 200 times the atmospheric pressure). However, in this case, the energy density per liter is much lesser than petrol, and transportation over long distance becomes expensive. In any case, the use of natural requires specialized infrastructure to be built which makes natural gas expensive. Secondly, the deposits of conventional natural gas are not as widely distributed as coal. Just like petroleum, a few countries command the majority of reserves. Iran, Qatar, Saudi Arabia, and UAE control about 37% of reserves, while Russia controls about 23% of reserves. Thus, gas will need to be transported over long distances for the use of natural gas to increase. Given these characteristics, it is possible to understand the reason that a transition to natural gas has not happened yet; although natural gas finds use in electricity generation, transportation, as well as for domestic applications such as cooking, water heating, and space heating.

With this information it is easy to understand why any one of the fossil fuels cannot replace others in the near future. Table 4.2 summarizes the share of each fossil fuel in the global TPE consumption for the years, 1990, 2000, and 2010 (*Sources: [EIA], [BP 2011]*).

Year	Petroleum	Natural Gas	Coal
1990	39%	22%	26%
2000	39%	24%	24%
2010	34%	24%	30%

Table 4.2: Fossil Fuels: Share of Total Primary Energy Consumption

How Much Fossil Fuel Energy Does $100 Buy?

It is quite interesting to compare coal, natural gas, and petroleum, in terms of the energy that $100 buys when used to purchase these fossil fuels. It is also useful to assess the impact of imports. Table 4.3 shows the prices of these commodities in the year 2010 and calculates the energy that $100 buys. The price data in this table is sourced from the US EIA except the data for Japan, which is sourced from the BP Review of World of Energy, 2011 [BP 2011]. The calculation of energy content in this table is based on the data specified in Table 2.5, Energy Content of Fuels. From this

Commodity	Price	Energy: GJ/$100
Coal: US sub-bituminous	$67 per metric tonne	43.28
Crude Oil: Western Texas Intermediate (WTI)	$79 per bbl	7.45
Natural Gas: US Henry Hub	$4.37 per million BTU	24.14
Coal Import: Japan (12,500 BTU) cif	$105 per metric tonne	27.62
Crude Oil Import: OECD Average	$84 per bbl	7.00
Natural Gas (LNG) Import: Japan cif	$10.91 per million BTU	9.67

Note: cif is cost + insurance + freight

Table 4.3: How Much Energy Does $100 Buy?

table, it is easy to understand the reason that coal still supplies a significant portion of the global energy consumption. Even imported coal may provide more energy per $100 as compared to natural gas from some domestic sources. Crude oil provides the least energy output per $100, yet natural gas has failed to replace it. The convenience of petroleum based liquid fuels (petrol, diesel etc.) is hard to beat. The data in this table is for 2010. Since then, crude oil has remained at a very high price and natural gas prices have fallen. In January 2012, WTI crude sold at a spot price of $99 while natural gas was at $2.5 per million BTU. This implies that crude oil returns only about 6 GJ per $100 and natural gas returns about 42 GJ per $100. Yet, a rapid transition from crude oil to natural gas as the fuel for transportation is not on the cards. Such a situation is certainly a cause for worry since it underlines the fact that petroleum is irreplaceable. If petroleum were to run out, the entire civilization will come to a standstill! The important question is, will the petroleum run out in the near future? There are two schools of thought in this regard: The Cornucopian Theory and the Peak Oil Theory.

The Cornucopian Theory

Every year, oil reserves are reported by the oil producing nations. The reported numbers are sometimes mistakenly interpreted to represent the amount of oil in the earth's crust. Such an interpretation is incorrect and neither is it necessary to know exactly how much oil is present in the crust. According to the cornucopian theory, what really matters is the amount of oil which can be extracted economically. For a given market price of oil, only certain deposits of oil can be exploited profitably. Only such deposits can be classified as reserves. As the market price of oil rises, more and more deposits may become profitable for extraction, assuming that the cost of exploration and extraction does not increase by the same factor. In such a situation, there will be a rise in the oil reserves, with an increase in the price of oil. Thus, reserves of any mineral are a function of market price and the exploration/extraction costs. This is the basis for the cornucopian economics, which states that we will never reach a point where the earth's crust will run out of a mineral. Far before such a point is reached, the price of the mineral will rise due to the reduced supply. Such a price-rise will either trigger more exploration of the mineral along with the development of technology to make the exploration/extraction cost-effective. Alternatively, the price-rise may trigger a substitution by some other product, which satisfies all the requirements. In either case, the demands of mankind will always be satisfied. The cornucopian theory assumes that human effort and innovation will always succeed in meeting the demands of the civilization. In this case, it is a moot question to ask how much oil is really there!

The Peak Oil Theory

At the other end of the spectrum are the experts who warn that the end of the industrial civilization is close, as earth is running out of petroleum. They argue that a peak in petroleum production will occur very soon, or has occurred already. The peak oil theory is based on the initial work by US geologist, M. King Hubbert. He stated that the extraction of any non-renewable resource follows a bell shaped curve. Initially, the resource extraction rises exponentially. Extraction reaches a peak when half of the resource has been extracted. After that, the production of the resource declines sharply and permanently. This theory was successfully used to predict the peak for the US petroleum production in 1970. Illustration 4.2 shows the US production of crude oil from 1900 to 2010. A peak in the production is clearly seen in the year 1970, as predicted by Hubbert's theory. The proponents of the peak oil theory propose different time-lines for the peak in the world petroleum production; but almost all agree that the peak will occur sometime before the year 2020. They argue that the reserves published by many oil producing nations are a sham. Most of the major oil exporting nations are autocratic regimes and the oil reserves reported by them are inflated for various

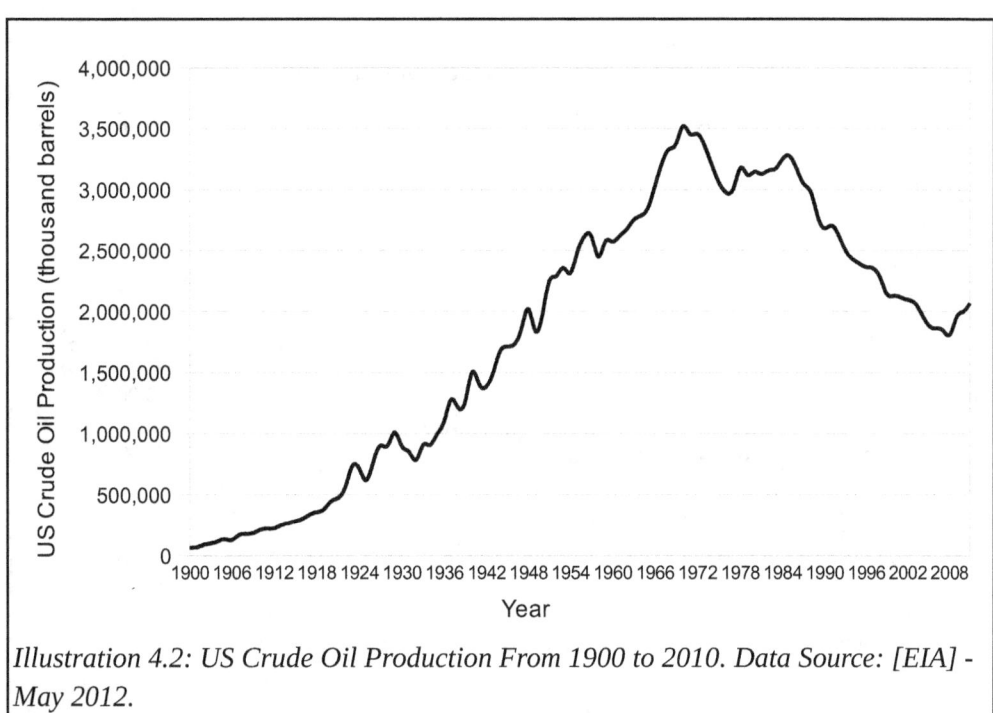

Illustration 4.2: US Crude Oil Production From 1900 to 2010. Data Source: [EIA] - May 2012.

reasons. There is also an argument that most of the newly discovered oil fields are either small, or expensive to extract oil from. The quality of newly discovered oil is poor, requiring expensive refineries. Most of the cheap oil that was to be discovered has already been discovered. Secondly, the output from many of the old oil fields may be on the decline. Moreover, the non-conventional resources, such as the Canadian oil sands and oil shales, though very large in quantity, are spread out over a vast region. Extraction of oil from such resources is not only expensive but there is also a severe risk of environmental damage. All these factors combined put a limit on how much oil the world can produce. In summary, the peak-oil camp predicts an impending crisis due to the expected fall in the petroleum production. The price-rise of crude oil since the year 1998 could be an indication of the approaching peak in production.

How Much Oil Remains?

This is certainly a very important question to answer. In chapter 3, the linkage between energy, economy, and human development was established. Earlier in this chapter, it was discussed that the world is not weaning away from oil, in spite of the high prices that have resulted in a very low energy return as compared to natural gas

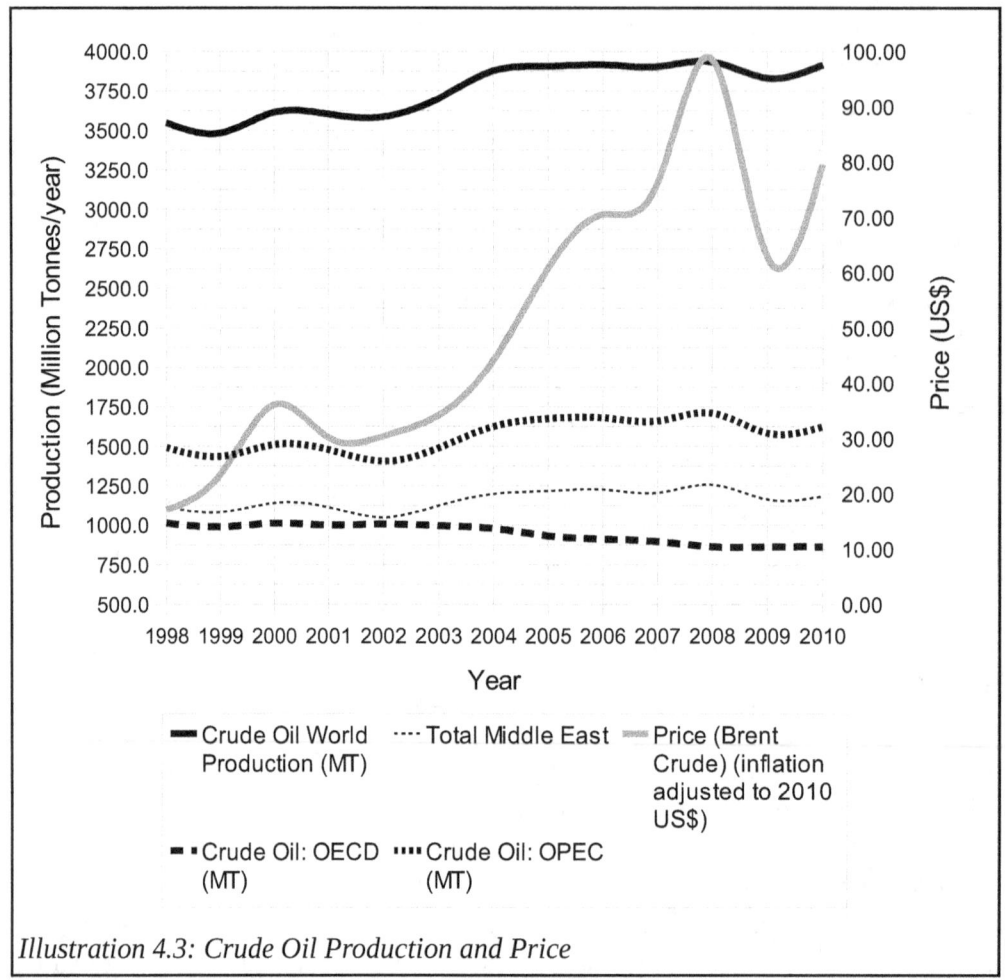

Illustration 4.3: Crude Oil Production and Price

and coal. Given these two facts, peak oil can certainly be a disruptive phenomenon. Reduction in the supply of oil will certainly lead to an economic crisis as all economies will face the prospect of deflation. Another consequence of peak oil will be the impact on human development. Gains in human development over the last fifty years or more may be wiped out as a result of peak oil. Given the seriousness of the consequences of peak oil, it is necessary to examine the arguments from the cornucopian and the peak-oil camps in detail. It will be useful to consider some hard facts:

A. Price Rise and Petroleum Production

Illustration 4.3, Crude Oil Production and Price, shows some very interesting data. Since 1998, the price of crude oil has been on a steep rise. The average yearly price of Brent crude increased from about $17 in 1998 to $99 in 2008 in terms of US dollars, inflation adjusted to the year 2010. This price reduced slightly to about $80 in 2010 and continued to rise in 2011. The rise from 1998 to 2010 works out to be 370%. These are the average prices for the year and there were some sharp, short-lived increases and falls during the 2008-09 period. Illustration 4.3 also shows the petroleum production from 1998 to 2010. It can be observed that petroleum production in the OECD nations, which are the main consumers, actually reduced by about 12%. Oil production by OPEC increased by less than 9%, while the total production in the world increased by only 10.3% during this 13 year interval. Thus, the rise in the production of crude oil was negligible compared to the price rise, even after factoring in the inflation. This data warrants some observations:

1. According to the cornucopian theory, a high price of crude oil not only triggers the exploration of hard and non-conventional sources of Petroleum but may also encourage substitution by natural gas as well as renewable energy technologies. In this situation, it is intriguing that OPEC nations did not increase the crude oil production to help stabilize the price. Increasing production would also have increased the OPEC market share, thereby increasing the OPEC's hold on the market. A high price of crude oil rattles economies around the world and could well be one of the causes of the recession in 2008-09. Such a scenario also impacts the investments made by OPEC nations throughout the world. Instead, an increase in oil production would have stabilized the oil price, generated good revenues, and helped the global economy during the period of recession. The most intriguing part is that during this interval, the proved petroleum reserves, as claimed by OPEC nations, increased from 823 billion barrels to 1068 billion barrels! Given that there is no rationale for not increasing the production to limit the price rise over this long interval of 13 years, there is only one conclusion that can be drawn: the OPEC nations simply did not have any spare capacity to increase the production of crude oil and they wanted the petroleum price to be high to be able to profitably exploit the reserves that they claim to possess. In other words, the OPEC nations simply cannot increase the production of conventional, light crude oil any

more. The oil reserves that they advertise are either a sham or consist of heavy oil which can be extracted profitably only at a high price.

2. The rise in the price of crude oil was stopped abruptly in 2008 due to the worldwide recession. Oil prices were below $40 for a few days. The prices quickly bounced back as the governments around the world took measures to reverse the economic situation. Thus, the price rise was halted only due to an anticipated fall in consumption and not due to an increased supply as the cornucopian theory suggests.

B. Non-conventional Petroleum Sources

Canadian oil sands are a major non-conventional oil source. Resources in Venezuela could be even larger and consist of heavy and ultra-heavy oil along with oil sands. Together, the resources in Canada and Venezuela are more than double that of the reserves reported by Saudi Arabia. The problem in characterizing these resources to be reserves is that they require expensive infrastructure before useful liquids (petrol, diesel etc.) can be produced. The infrastructure includes specialized oil-wells as well as refineries. Secondly, the processing required is quite energy-intensive. Only at a high market price it is economical to exploit these resources. Another problem is the environmental concerns associated with extraction and processing of such resources. Hence, ramping up production from non-conventional resources to replace the conventional crude oil on a large scale is difficult. During the 1998-2010 interval, when oil prices rose by 370%, Canada reported an increase of 30% in the production of crude oil, thanks to the increased output from oil sands. However, as a percentage of the total demand, this increase in output was too little to impact the price of crude oil. Output from Venezuela actually declined during this interval from 179.6 MT in 1998 to only 126.6 MT in 2010 [BP 2011]. This data is an indication of the fact that while the non-conventional petroleum resources will be available for a long time, their ability to ramp up production on a scale to replace the conventional petroleum is doubtful.

C. Cornucopian theory and Thermodynamics

The cornucopian theory is an economic theory. Any hypothesis that derives itself from economics cannot violate the more fundamental laws of thermodynamics. As the resources of a mineral deplete, production falls, leading to a rise in price. The increased price makes it economical to exploit the sparse and hard-to-access deposits of the mineral. While this is true, the exploitation of such deposits requires more energy. However, if the mineral being extracted is itself a source of energy then there is a problem! The amount of energy required to mine and process the mineral should be less than the energy content of the mineral. For instance, the energy content of crude oil is about 37 MJ/liter. The energy required to mine and process crude oil, also called the energy input, must be less than 37 MJ/liter. In other words, the EROEI for extraction, exploration, and delivery of the mineral should be greater than 1. It is not easy to measure the actual energy input. Apart from the energy actually spent on drilling and processing, it is necessary to take into account the entire economic activity related to exploration, administration, delivery etc. The first law of thermodynamics states that energy cannot be created or destroyed. The second law of thermodynamics states that unavailable energy (entropy) always increases. A conclusion from these laws of thermodynamics is that there will be a point when extraction of an energy resource will not be useful as there will be a net loss in energy. It may be argued that the extraction of the resource may still be useful if some other cheap energy source is available. For instance, some of the processes that are a part of extraction of non-conventional petroleum may use natural gas as the energy source. Such an extraction of petroleum may be useful on a small scale but not on a global scale, since the demand for natural gas will then skyrocket and the problem will soon transfer to natural gas.

D. Peak Oil and Impact of Variables

The peak oil theory that predicts the bell-shaped curve for extraction of a resource assumes that there are no other variables that impact the resource extraction. Same assumption is made when the peak oil theory is applied to the global production of petroleum. However, for a product that virtually drives the global economy, such an assumption will not work. Oil price was one of the important factors that contributed to the global recession in 2008-09. A global recession is bound to impact demand, which will bring the petroleum price down. Secondly, volatility in petroleum price can also affect investments in the petroleum sector. Finally, high petroleum prices trigger the efforts to harness non-conventional petroleum sources as well as substitutions by other/renewable sources of energy. These variables imply that a peak in production followed by a rapid decline is not what will likely happen. Instead, a high long-term average price of petroleum, along with high volatility in the short-term prices is the most likely outcome. As far as the production is concerned, it may reach a plateau rather than a peak.

E. Energy Transitions

The cornucopian theory assumes that a high price of a mineral will make substitutions by other minerals more viable. Applying this theory in the case of petroleum, a transition to alternate sources of energy must take place, considering the rise in price of petroleum since 1998. The only viable fossil fuel source for such a transition is natural gas. It was calculated earlier in this chapter that natural gas returns about 7 times the energy per dollar, as compared to petroleum. However, there is no evidence of a major transition taking place. Secondly, natural gas has a rather large burden to carry. Apart from the current applications, here are some applications which need to use natural gas:

1. Replacing petroleum as a transportation fuel.
2. Replacing coal for electricity generation in most parts of the world.
3. Helping in the extraction of non-conventional petroleum sources.
4. Replacing kerosene and even LPG for household cooking applications in countries like India.

It is necessary to examine if natural gas resources can really meet the increase in demand due to these applications. The data in the 2000-2010 decade shows that the share of natural gas in the global primary energy consumption has remained stagnant at 24%. During this interval, the consumption of natural gas, coal, and petroleum continued to increase, when measured in absolute terms. Consumption of coal grew at the fastest rate (60%), followed by the growth in natural gas consumption (32%), and the petroleum consumption (14%). This data certainly does not point to any imminent transition at a global level. Even in USA, in the 2000-2010 decade, the share of natural gas in the total primary energy consumption increased only marginally, from 23% to 25%. During the same interval, the share of petroleum consumption dropped only marginally, from 39% to 37%. Once again, hardly an evidence of transition. As a result, in January 2012, the price of petroleum remains high while the price of natural gas remains at a multi-year low value.

It was stated earlier that the gaseous nature makes it difficult to store and transport natural gas. Compressed natural gas has a poor energy density as compared to petrol, and automobiles retrofitted to use CNG have to sacrifice trunk space to house the bulky CNG cylinders. Shipping liquefied natural gas requires specialized cryogenic tankers as well as storage terminals. Natural gas can also be transported using pipelines. However, this too is an expensive option and connectivity to a pipeline cannot be as ubiquitous as the electric grid. In conclusion, a transition to natural gas,

Illustration 4.4: A CNG Bus at the Denver International Airport, USA. Courtesy of DOE/NREL, Credit - Dean Armstrong

simply due to the high price of petroleum, is not very likely. It may require several decades to build the infrastructure for such a transition to occur.

This section started off with the question, "How much oil remains?" After all the discussion in this section it must be concluded that the question itself needs to be changed to, "At what price and cost will the oil be available?" Quite surely, earth still has a lot of oil, when the non-conventional sources are factored in. The question is about the price at which it will be economical to extract petroleum at a scale to meet the global demand. We don't know if peak oil has already occurred for the conventional sources. Neither will any organization (such as OPEC) will declare a peak. What we know is that the oil price certainly bottomed out in 1998. The high price of petroleum has failed to increase production and could be an indicator of a peak in production for the conventional sources. The oil production may increase in the future, thanks to the non-conventional resources. However, the projections for demand in the year 2030 are rather huge, and producing petroleum to meet this demand at a reasonable price, in an ecologically safe manner, does not seem possible. As a result, it will be wrong to depend on petroleum to be the engine for world economic growth. Practically, earth will never *run out* of petroleum. The accessible resources will just get progressively expensive until the demand tapers off. Some resources will always remain unaccessible due to the cost of extraction, ecological costs, and due to excessive energy input for extraction.

Natural Gas Reserves

Reported reserves of conventional natural gas are about the same as that of petroleum, at about 180 billion tonnes. For natural gas, however, there does not seem to be any imminent problem on the supply side. The reserves are concentrated in a few nations, and the current rate of production indicates that they will last for more than a century for some of the producers. For instance, Iran has 15.8% and Qatar boasts of 13.5% of global reserves. For both these nations, the R/P ratio is higher than 100, implying that the reserves will last more than a century. Russian Federation has 23.9% of the global reserves with an R/P ratio of 76. By the way, these numbers are for conventional, or easy-to-obtain natural gas. Deposits of non-conventional sources are not even accounted for these nations. USA is the main consumer of natural gas and R/P ratio for USA is just 12.6. This is in spite of addition to reserves due to the success in extraction of shale gas, a non-conventional resource. Even the OECD group of nations, which include USA, have an R/P ratio of just 14.7. China and India, which have a high growth in energy consumption, have an R/P ratio of about 29. Though this is a reasonable number, China and India rely on coal as the primary source of energy. Any transition from coal to natural gas is certain to deplete the conventional natural gas reserves rapidly in these countries. Thus, the main consumers of natural gas; viz. OECD nations, China, and India, have very modest reserves that are depleting quickly. At the same time, a few countries have huge reserves that are not being consumed very rapidly. Once again, this points to the difficulties in transition to natural gas. As a result, the global price of natural gas remains quite low. Another important data point is related to the percentage of exports as compared to production. In 2010, out of a total production of 2881 million tonnes of natural gas, only about 30% was exported. In fact, exports of LNG were only 268 million tonnes or only about 9%. During the same year, out of a petroleum production of 3913 million tonnes, 1875 million tonnes, or 48% was exported [BP 2011]. Clearly, the import and export of natural gas has limitations. As a result, unlike petroleum, depletion of reserves of natural gas is a localized problem and not a global one.

Natural Gas Usage and Resources in USA

In January 2012, in his state of the union address, US president Obama announced that US now has a 100 year supply of natural gas, thanks to some technological advancements. One country that was facing the problem of depletion of natural gas resources was (or perhaps still is), USA. This country has the highest consumption of natural gas. Consider the following statistics for USA for the year 2010 [BP 2011] :

- Consumes about 22% of the natural gas produced in the world.
- Natural gas production of 557 MTOE, the highest in the world.
- Possesses only about 4% of the natural gas reserves in the world.
- A poor R/P ratio of 12.6, which implies that USA will run out of natural gas in about 13 years.

The statistics of 2010 is actually better than the year 2000, thanks to the new technology that makes it possible to tap the shale gas resources. This is one of the striking demonstrations of the cornucopian theory. Faced with the depletion of conventional natural gas, fracking technology was developed to harness the non-conventional shale gas resources.

Shale Gas and Fracking

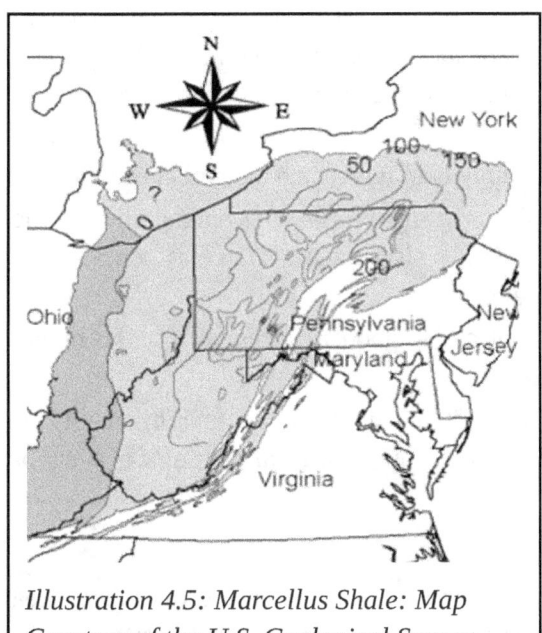

Illustration 4.5: Marcellus Shale: Map Courtesy of the U.S. Geological Survey

Shale gas is the term used for natural gas harnessed from shale rocks, a type of sedimentary rock. Shale rock formations useful for natural gas production are located deep in the earth's crust and span a large area. For instance, the Marcellus Shale in the eastern USA spans parts of Ohio, Pennsylvania, Virginia, and New York. This shale formation is located about 700 to 2500 meters deep. The thickness of the shale formation is just a few tens of meters. Shale rock is a porous rock with low permeability. Gas and oil is generated due to organic matter trapped inside such rocks.

Due to the low permeability of the rock, most of the oil and gas remains trapped inside the pores. There is almost no accumulation of oil and gas that would allow for easy extraction. Thus, useful oil and gas is spread over a large geographic area with a low yield per square km. The usual, vertically drilled natural gas wells, do not yield much output due to the absence of accumulation of natural gas in a natural reservoir. The fracking technology attempts to solve this problem by forcing accumulation in an artificial way. In this case, drilling is carried out vertically until the shale formation in the crust is reached. Now the drilling proceeds horizontally within the shale formation (see Illustration 4.6). The horizontal section of a well can be more than a kilometer in length. Once such horizontal sections are drilled, a mixture of water, sand, and some chemicals is injected at high pressure in the well. This causes the shale rock to develop fractures. Natural gas and oil is released, as pores in the shale rock open and fractures allow them to accumulate in the well. Such a well now can now produce gas and oil.

Shale Gas: Pros and Cons

Shale gas production is dependent upon the fracking technology which has been quite controversial. There have been arguments from both sides:

Pro Fracking and Shale Gas:
- Tapping the shale gas resource can potentially provide an alternative source of energy which will reduce the dependence on petroleum. Shale gas has certainly helped US ramp up the otherwise declining production of natural gas from a low of 468 MTOE in 2005 to 557 MTOE in 2010.
- A vast new energy source will certainly help rejuvenate the global economy.
- There are vast deposits of natural gas in shale formations spread all over the world. This is certainly a democratically distributed resource as compared to petroleum and conventional natural gas.
- Natural gas is a clean-burning fuel and a good replacement for the highly polluting coal.
- Burning natural gas results in less CO_2 emissions per unit of energy as compared to all fossil fuels. This is very important from the point of view of global warming.

Against Fracking and Shale Gas:
- Shale gas fields (called 'plays') are characterized by a spread over large area and a low energy density. For instance, for the Barnett shale in Texas, the

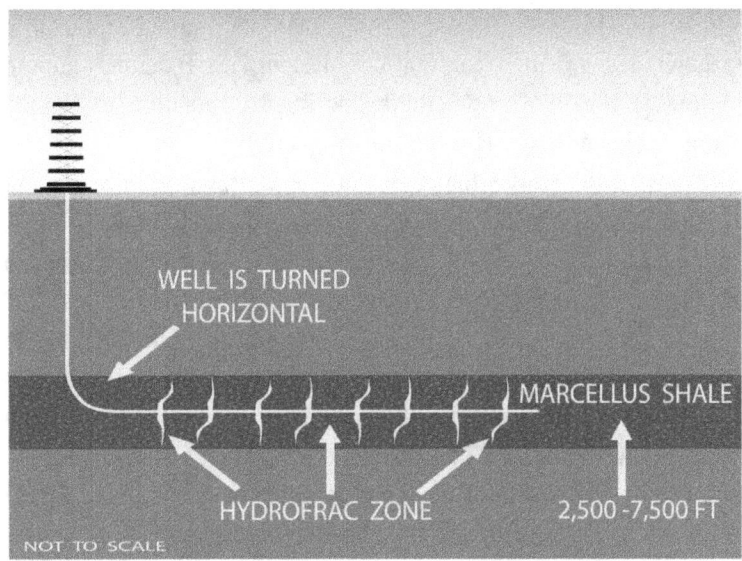

Illustration 4.6: Horizontal Drilling and Fracking the Shale Rock. Credits: Laurie Barr/Shutterstock.com

energy content in shale rock is only about 1000 to 1400 Btu/cubic-foot. The energy density from the perspective of land area is 50 to 250 Billion-cubic-feet(Bcf)/square-mile [MIT 2011]. These numbers convert into 0.037 to 0.052 MJ/liter and $21 - 104$ GJ/m^2 respectively, in metric units. Not all of this gas can be recovered. The recovery factor is generally about 0.25. This means that the energy available per square meter is comparable to solar insolation, which is about 7 GJ/m^2/year, depending on the location. While solar energy is available forever, the shale gas in a producing region would be consumed in just a few years and is then gone for good. Secondly, any energy source with a low energy density generally requires a high capital cost and a large upfront energy input. Such energy sources deliver comparatively lower net energy, implying a lower EROEI.

• There is a fear that the process of fracking may lead to contamination of fresh-water sources. Indeed, cases have emerged in USA where fracking seems to have contaminated local water supply at some locations (see box). In 2005, fracking was exempted from the Clean Water Act. As a result, the US Environment Protection Agency (EPA) did not have the authority to regulate fracking. This was perhaps alright in the initial phase of research and development. However, now that more than one hundred thousand such wells have been dug up all over the US, having no federal regulation is quite suspicious. Even in Europe, there has been opposition to fracking. In UK, fracking has been linked to even seismic activity.

- One important difference between the conventional natural gas and the shale gas is that an awfully large number of shale gas wells need to be drilled. On the average, shale gas wells produce much less gas than the conventional wells. Production from shale gas wells is known to decline by more than 70% during the very first year of operation. According to the analysis presented in an MIT study [MIT 2011], mean output from wells in the Barnett and Marcellus shales after one year of operation reduced to less than 1 million cubic feet (MCF) per day. Thus, to keep the gas flowing, drilling activity must continue. The daily natural gas consumption in US is about 66,100 MCF. In order to satisfy 60% of this demand with shale gas, assuming a stable output of 0.5 MCF per day per well, about 79,000 shale gas wells need to be drilled. At a cost of about 4 million dollars per well, this requires a capital investment of 316 billion dollars. The question that needs to be asked now is if the shale gas makes economic sense and at what market price. The answer to this question depends on how long a well will continue to operate on the average and what is the expected ultimate recovery (EUR) from an average well. The price fetched at the head of the well needs to offset the operational expenses, depreciation, and provide a reasonable return on the capital invested. Analysis by MIT [MIT 2011] indicates that there is a wide variation in the break-even price across various shale fields as well as within a given shale. The variation is from as low as $2.88 to $17.04 per TCF (Thousand Cubic Feet). Clearly, it is too soon to determine if shale gas makes economic sense and for which plays. In January 2012, natural gas prices are below $3 and at this price almost all shale plays, except those that also output oil or other associated hydrocarbon liquids (NGLs), are expected to be loss making.

- It is necessary to ascertain if shale gas returns any net energy and saves on CO_2 emissions. Drilling, fracking and subsequent cleanup are energy-intensive operations. A single well requires more than a million gallons of water mixed with sand and chemicals. The so called 'produced water', or toxic waste water in other words, must be disposed of safely to avoid environmental damage. Developing each shale gas well requires about $4 million. This spending not only corresponds to energy consumption but also to CO_2

emissions. Secondly, natural gas needs to be processed and pipelines need to be built to connect the wells to a distribution network. These activities involve operational costs and capital costs. It is also necessary to take into account the cost of capital. All these costs translate directly and indirectly into energy consumption and CO_2 emissions. The wells so built decline in output very rapidly and have a low EUR. For instance, wells in the Barnett shale in Texas have an EUR of about 2.4 BCF. It is not without simplifying assumptions that one can even attempt to calculate EROEI for shale gas. As a first step, it is simplifying to ignoring the cost of financing as well as the operational costs. Both these actually constitute an energy input. Another simplifying assumption would be that the capital cost involved in drilling a shale gas well directly corresponds to an energy input in the form of natural gas. Ideally, one would use the energy intensity of economy to calculate the energy input. However, the fracking activity is an extremely energy intensive activity, while the economy is dominated by the service sector which uses very less energy per dollar of economic output. With these assumptions, and a price of $4 per TCF of natural gas, the $4 million cost of a shale gas well translates to 1 BCF of natural gas. This is the energy input. Output (EUR) from the shale gas well is expected to be about 2.4 BCF. Thus, EROEI is calculated to be 2.4, which is a very low number. Obviously, this calculation is very simplistic but illustrates the point well that EROEI for shale gas is unlikely to be very good.

US Shale Gas: 100 Year Supply?

Now that the pros and cons of shale gas have been discussed, it will be interesting to evaluate if US indeed has a 100 year supply of shale gas. There is a wide variation in the estimates of the technically recoverable gas that is available in the earth's crust. For instance, the United States Geological Survey (USGS) estimates the shale gas resources to be 336 Trillion Cubic Feet (TRCF) [USGS 2011], while the Petroleum Guidance Group (PGC) estimates this number to be 616 TRCF [MIT 2011]. It is very difficult to reconcile the large variation in these numbers. Secondly, these are technically recoverable resources that may not be economical or may not even deliver any net energy. The MIT report [MIT 2011] uses a methodology that all technically recoverable resources can be extracted provided that the market price is sufficient to recover the associated costs. This methodology has a couple of problems: Firstly, at a high market price, demand may get decimated and any shale gas recoverable only at that price will remain in the ground. Secondly, there is an assumption that the production cost is independent of the market price of natural gas. This is an error since rising energy prices will impact the production cost. In fact, it is hard to imagine that natural gas prices will increase on a stand-alone basis. It is expected that there will also be a corresponding rise in the petroleum prices which has always been the preferred fuel. A rise in the price of natural gas will also lead to an increased demand for coal, leading to a price rise. Such a simultaneous rise in the price of fossil fuels is bound to impact the cost of production. Hence, it is impossible to put a number on the quantity of shale gas that can actually be recovered. Assuming just for the sake of argument that the USGS number of 336 TCF also represents the amount of shale gas that can be recovered economically, it represents a 14 year supply based on the natural gas consumption in 2010.

Coal Reserves

Coal is the most democratically distributed fossil fuel resource. It is also the cheapest fossil fuel, providing more energy output per dollar of expenditure (see Table 4.3). As a result, coal is still a major contributor to the world energy consumption, despite the harmful environmental impact that ranges from CO_2 emission to the contamination of water bodies with sulfuric acid, and from mercury pollution to radioactivity. According to the BP Statistical Review of World Energy, 2011, the R/P ratio for coal is 118 for the world reserves. In other words, at the current level of consumption, the currently discovered coal deposits would last for 118 years! For countries with high consumption of coal, the R/P ratios are the following [BP 2011]:

- USA - 241,
- India - 106,
- Russia - 495,
- China - 35.

Barring China, these R/P ratios are very comfortable for the ongoing consumption, as well as for the growth associated with the expansion of economy. There is also reason to believe that with abundant reserves of coal, the hard-to-access coal deposits are not explored yet.

Coal Consumption in India

Apart from hydroelectricity, coal is the cheapest source of electricity production. In the previous two decades, coal consumption in China increased at a very high rate. It is possible that the same may happen in India in the coming future, given the rapid

Illustration 4.7: An Open-Pit Coal Mine. Credits:Straga/Shutterstock.com

economic growth. In fact, in the year 2000, India produced 335 million tonnes of coal, while in the year 2010, coal production was at 570 million tonnes. This rise in production could not meet the domestic demand, and India had to import 88 million tonnes of coal in 2010 [IEA 2011]. The coal consumption is likely to accelerate further since there is no cheaper alternative for generating electricity. Earlier in this chapter (see Table 4.1), it was estimated that the coal consumption in India will rise to 3,300 million tonnes in the year 2030. There are a number of challenges in producing so much coal. There are environmental hazards of coal mining and acquisition of land for coal mining is a major problem. There is a rising anger against displacing people since compensation is often insufficient and delayed. Environmental groups are opposing the clearing of forests for coal mining. Finally, the quality of coal is also an issue. The coal produced in India generally has a lower calorific value and a higher ash content as compared to the coal imported from Australia and Indonesia. These problems partly explain the reason why India had to resort to coal imports even in 2010, despite having huge reserves. Coal reserves in India were reported to be 60,600 million tonnes in 2010 [BP 2011]. At the present rate of consumption, these reserves will last for about 106 years. If the projected consumption rate of 3,300 million tonnes per year is actually achieved in 2030, the coal reserves in India would last only for about 20 years. This calculation assumes that new discoveries and revaluation of existing coal fields balance the depletion due to ongoing consumption until 2030, and reserves are maintained at 60,600 million tonnes. In spite of such a high projected consumption, there is an expected shortfall in the production of electricity. In another estimate that related to the shortfall in the production of electricity, it was estimated that India may need to import almost 1.6 billion tonnes of coal annually in the year 2050 [Kakodkar 2009]. This is a truly mind boggling number which is unlikely to be satisfied in the era of depleting fossil fuel sources. Even today, in the year 2012, importing coal is proving to be a major problem. International price of coal has risen significantly in the past few years and exporting nations such as Indonesia have changed laws to control the export of coal. Such a change has proven to be debilitating to the large power projects that are dependent on imported coal [Jayakumar 2012].

Fossil Fuels in the Future

The discussion in this chapter should make it clear that fossil fuels will continue to be available in the future and any talk of running out of fossil fuels is certainly an exaggeration. However, making our future dependent upon the ample availability of fossil fuels at low prices is certainly a folly. Market prices of fossil fuels are likely to remain at high levels. The linkage between economy and energy consumption was discussed in chapter 3. A reduction in energy consumption due to the high prices of fossil fuels will lead to a recessionary economic situation. The developing nations, which must achieve a high economic growth rate to meet certain social objectives, will not be able to depend upon the use of petroleum and natural gas as the fuels for growth. The developed nations too will need to face an impact on economy due to the high price of crude oil while a transition to natural gas to replace petroleum does not seem very likely. Another problem could be the possible increase in the consumption of coal that will lead to environmental and social problems. In summary, the era marked by the use of fossil fuels as the engines of growth is very close to being over, though they will continue to be available for the foreseeable future. This situation can either be viewed as a crisis or as an opportunity since a strong push for the renewable sources of energy may reduce the dependence on fossil fuels while creating a new engine for economic growth. The next chapter covers the renewable sources of energy in detail.

5. RENEWABLE SOURCES OF ENERGY

Before the industrial revolution, mankind was dependent on biomass for satisfying all the energy needs. Biomass mostly consisted of firewood, crop residues, cow dung, animal fat, and charcoal. Biomass is essentially a renewable resource as long as the use is in equilibrium with the process of generation in nature. An energy source is classified as renewable if the energy supply from the source is practically permanent. Such sources are also called the *green* sources of energy. A common characteristic of all renewable sources of energy is that they derive their energy from the sunlight that reaches the earth[1]. The following are the major renewable sources:

- Hydroelectricity
- Solar Photovoltaic Electricity
- Solar Thermal Energy
- Wind Energy
- Energy from Biomass and Biofuels

Tidal energy and geothermal energy are also renewable sources with some potential. However, these sources won't be covered in this book since the promise that they hold is not as established or as widespread as the other sources. All other renewable sources will be analyzed in this chapter to ascertain the potential and to estimate important parameters such as, power/energy density and EROEI.

Hydroelectricity

Hydroelectricity is the oldest of the modern renewable sources of energy. Some of the largest hydroelectric projects were implemented over the last 50 years. According to

---

1 *Geothermal energy is an exception. It uses heat in the earth's core.*

the statistics published by the International Energy Agency [IEA 2011], in the year 1973, OECD nations produced about 72% of the hydroelectricity generated in the world. In the year 2009, this number reduced to only about 42%, as new hydroelectric projects were implemented in the developing nations. The three largest hydroelectric projects are all outside the OECD group of nations:

- Three Gorges Dam, 22 GW, China
- Itaipu Dam, 14 GW, Brazil and Paraguay
- Guri Dam, 10 GW, Venezuela

By any standards, these are huge generation capacities. There are 5 nations in which more than 40% of the domestic generation of electricity is based on hydroelectricity [IEA 2011]:

- Norway 95.7%
- Brazil 83.8%
- Venezuela 72.8%
- Canada 60.3%
- Sweden 48.3%

In spite of this impressive statistics, hydroelectricity supplies only about 6.46% of the TPE consumption of the world [BP 2011]. Secondly, in the year 2009, only 16% of the total electricity generated in the world was based on hydroelectricity, while 67% was supplied by fossil fuels [IEA 2011]. In the case of India, only about 5% of the TPE consumption is supplied by hydroelectricity while fossil fuels supply about 93% of the TPE consumption. There is a lot of value in analyzing the pros and cons of hydroelectricity along with the potential that it has to become the prime source of energy. The following paragraphs discuss various characteristics of hydroelectricity.

Is Hydroelectricity a Clean Source of Energy?

It is generally accepted that hydroelectricity is a clean source of energy. After all, it is just the flowing water that runs the turbines to generate electricity! No fuel is being burnt and neither are any radioactive fission products being created. Well, it is not quite as rosy as that:

A hydroelectric power station requires a large reservoir of water situated at a height. When a dam is built on a river in a suitable geography, an artificial reservoir is created. Water from such a reservoir inundates a large land area, submerging forests and other vegetation. This submerged organic matter decays over a long time, producing CO_2 and methane (CH_4), which are greenhouse gases. Reservoirs of large

Illustration 5.1: View of Guri Dam in Venezuela

dams in the world occupy more than 500,000 km^2 of area. Forest and vegetation in this area would normally be a part of carbon cycle: while the decay of dead vegetation would continue to release greenhouse gases, it will be offset by the absorption of CO_2 as vegetation and other dependent organisms and animals grow. Once the forest is submerged, the carbon cycle is broken and organic matter decays to produce greenhouse gases (GHG) over a long time. There have been studies which conclude that the GHG generation due to such artificial reservoirs per unit of electricity is close to that of a coal fired power station [Smil 2005]. However, jury is still out on this topic since accounting of GHG emissions and GHG savings over the hundred-year lifespan of a hydroelectric project is not simple.

Another important point is that building large dams impacts the downstream flow rate of water. This results in depletion of downstream wells and groundwater. Secondly, the dam and the reservoir block the flow of silt which enriches the soil. As a result, an impact is expected on the downstream carbon cycle as well. This and other factors, which are out of the scope of this book, imply that ecological damage is an inevitable consequence of building large dams.

Displacement of human settlements due to large dams is another major problem associated with hydroelectricity. There is a growing opposition to building large dams, especially in the areas full of biodiversity and at locations where a large number of people will be displaced. The movements to oppose the Sardar Sarovar dam and the Tehri projects in India were very strong. Even in a democratic country like India, compensation to displaced individuals is rarely adequate or timely, and is mired in red tape and bureaucracy.

Power Density and Energy Density for Hydroelectricity

Power density is calculated by dividing the maximum power output measured in Watts by the surface area of the reservoir. Energy density is calculated by dividing the average yearly energy output measured in kWh by the surface area of the reservoir. The power density is measured in W/m^2 while the energy density is measured in $kWh/m^2/year$. It is desirable that hydroelectric projects have a high power density and energy density since such a number indicates a better utilization of land. These two numbers also serve as benchmarks to compare the different sources of energy.

Power density and energy density of hydroelectric power projects varies significantly, depending on the geography of the location. In general, projects located high up in the mountains provide a better density as compared to the projects closer to the sea level. In India, the Koyna hydroelectric power station in the Sahyadri mountains in Maharashtra has a reservoir of about 900 km^2 surface area with a maximum power output of 1920 MW [Koyna]. The Tehri project, located in the foothills of Himalayas, has a reservoir that spans 42 km^2 with a maximum power output of 1000 MW [Tehri]. Power density for Koyna works out to be 2.1 W/m^2 while that for Tehri is 24 W/m^2, an order of magnitude higher! For the Sardar Sarovar project on the Narmada river, the reservoir area is about 370 km^2 while the power generation capacity is 1450 MW [Sardar Sarovar]. The power density for the Sardar Sarovar project works out to be about 4 W/m^2. The Grand Coulee dam in USA has a reservoir of 333 km^2 with a power generation capacity of 6809 MW [Grand Coulee 2012]. The power density for the Grand Coulee dam works out to be about 20 W/m^2.

Hydroelectricity is a very inexpensive source of energy and its dispatchable nature makes it an ideal source of electricity. Hydroelectric power stations are often utilized to supply power during the period of peak loads on the grid since it is easy to turn them on and off at a short notice. They are sometimes used to supply base loads on the grid too, depending on the availability of power on the grid and the demand for electricity. There may be multiple uses for the water stored in the reservoir and it may not always be possible to operate the hydroelectric power stations on a continuous basis. As a result, most hydroelectric power stations operate at a low capacity factor. In the year 2005, the Koyna hydroelectric power station in India was operated at a capacity factor of just 18%. The Grand Coulee power station generates about 21 billion kWh of electricity per year, implying a capacity factor of 35%. The Tehri project operated at a capacity factor of about 45%, while the Sardar Sarovar project operated at about 33% capacity factor. With this data, it is easy to calculate the annual generation of electricity and the energy density. Table 5.1 summarizes the power density and energy density of these hydroelectric projects. It can be observed that while there is an order of magnitude difference among the hydroelectric projects, the power and energy density is extremely low. The Tehri project operates at an energy density of 76.6 kWh/m^2/year. Compare this with the solar insolation of about 1,825 kWh/m^2/year on an average location on earth. From this data it can be concluded that as far as the land use for electricity generation is concerned, hydroelectric power is not very efficient.

Project	Capacity Factor	Power Density	Energy Density
Koyna	18%	2.1 W/m^2	3.5 kWh/m^2/year
Tehri	45%	24 W/m^2	76.6 kWh/m^2/year
Sardar Sarovar	33%	4 W/m^2	11.5 kWh/m^2/year
Grand Coulee	35%	20 W/m^2	63.0 kWh/m^2/year
World Avg.	39%	3.9 W/m^2	13.1 kWh/m^2/year

Table 5.1: Power and Energy Density of Hydroelectric Power Projects

Positive Aspects of Hydroelectric Power

In spite of the negative aspects of hydroelectricity discussed here, it will be unfair to conclude that hydroelectric power is not a good source of energy:

One useful characteristic of hydroelectric power plants is the high value for EROEI. It is estimated that hydroelectric power plants will produce about 200 times the energy required to build, maintain, and operate the plant [Gagnon 2005]. This net energy return contributes to make hydroelectricity the cheapest source of electrical energy.

Another positive aspect is the obvious fact that there are no emissions other than due to the decay of the submerged vegetation. Life span of dams and reservoirs is generally very long, more than 50 years. During this long interval, almost no polluting gases or particulate matter are released into the atmosphere, making it a very clean source of energy.

Hydroelectric power is a byproduct of building dams, which have other important uses. The reservoir may be used to provide drinking water, irrigation, and flood control. These applications, to some extent, negate the greenhouse gas emissions from the reservoir.

While it is true that the low power density of hydroelectricity leads to huge utilization of land, the actual construction work of the dam takes place on a fairly small area. For instance, the Grand Coulee dam in USA has a reservoir of 333 km^2 while the length of the dam is just 1,592 meters. For the same dam, the volume of the reservoir is about 11.5 billion cubic meters while the amount of concrete used in the construction of the dam is about 9 million cubic meters [Grand Coulee 2012], a significantly smaller number. The dam on a relatively small area serves as the means to create a huge concentrated source of energy. It is this characteristic that contributes to make hydroelectricity relatively inexpensive while creating an almost perfectly dispatchable source of electrical energy. Depending on the demand from the electric grid, hydroelectricity generation can be adapted in a matter of few seconds. A hydroelectric power plant can be operated at practically any capacity factor, depending on the requirements from the electric grid and the water available in the reservoir. This characteristic makes it an extremely flexible source of electricity.

EUF and EROEI Calculations for Hydroelectricity

There are various reasons due to which EUF and EROEI calculations are difficult for hydroelectric projects:

Electricity is only one of the benefits of constructing a dam. Other benefits include flood control, irrigation, and municipal water supply. It is not possible to assign a specific percentage of the total cost to electricity generation. Without knowing the cost, it is not possible to calculate the energy input for the hydroelectric project.

Hydroelectric power projects take long time to plan and to build. The total cost is spread over several years. Even if generating hydroelectricity was the sole purpose of the project, it is not easy to calculate the energy input on the basis of expenditure over a long interval.

The cost of hydroelectric projects varies significantly based on the geographic region. For instance, the cost of a project in the foothills of Himalayas will be significantly higher than a project that generates a similar amount of electricity in the plains. This difference will also imply a difference in the energy input. As a result, EUF and EROEI will differ significantly for such projects.

According to a certain estimate [Gagnon 2005], hydroelectric power plants produce 205 to 280 times the energy required to build, maintain, and operate the plant, assuming a lifespan of 100 years. In other words, EROEI for hydroelectricity is 205 to 280. Due to the reasons previously mentioned, the EROEI calculation is expected to be approximate and indicative only. Yet, EROEI for hydroelectricity is very large compared to all other renewable sources of energy. Even an order of magnitude error in the calculation of this number will not make much change in the fact that hydroelectricity provides the best energy return. Assuming an EROEI of 200 and an operational life of 100 years, in the very first year of operation, a hydroelectric project will return 2 times the energy consumed in building it. In other words, EUF for hydroelectricity is estimated to be 0.5.

Hydroelectric Generation Potential in India

According to the National Hydroelectric Power Corporation of India, the total hydroelectric potential in India for conventional projects is about 84,000 MW at a 60% capacity factor. The potential for micro, mini, and small hydroelectric systems is 6,780 MW [NHPC]. This potential implies a generation of 1.3 billion kWh of electricity per day, or about 1 kWh per person per day. Only about 20% of this potential is harnessed already. If the remaining potential is harnessed, the availability of electricity per capita will increase by about 45% from the current level. If the hydroelectric potential is harnessed completely, it will result in an annual availability of 475 billion kWh. This number corresponds to a primary energy availability of 5 EJ, which is about 6.7% of the projected TPE consumption in the year 2030; hardly an improvement over the 5% of TPE consumption that hydroelectricity provides for in the year 2010. Moreover, there is an assumption in this calculation that all the remaining capacity can be harnessed by the year 2030, a very optimistic assumption! There are some basic problems in harnessing the complete potential:

1. Most of the remaining potential is in the north and the north-eastern parts of the country, in the Ganges and the Bramhaputra basin. This region is

environmentally sensitive and prone to earthquakes. The terrain is near the foothills Himalaya and building dams in this region will not only face opposition from the environmentalists, but there will also be a number of engineering challenges.

2. Many of the proposed hydro projects are in the north-east, located in the states of Arunachal Pradesh and Sikkim. This region is far from population and industrial centers, and electricity will need to be carried over a long distance, resulting in transmission and distribution losses.

In summary, while there is a significant potential for hydroelectricity to grow from the current level, harnessing the potential is difficult and will not improve much on the share of the TPE consumption that hydroelectricity provides in the year 2010. As a result, at least for India, even a partial transition to hydroelectricity from fossil fuels is out of question.

World TPE Consumption and Hydroelectricity

In the year 2010, hydroelectricity had a share of just 6.46% of the world TPE consumption of 12,002 MTOE, while the share of fossil fuels was about 87% [BP 2011]. It will be interesting to calculate the hydroelectric capacity required to reduce the share of fossil fuels to 75% of the TPE consumption. This implies converting about 12% of TPE consumption to hydroelectricity. In other words, about 1440 MTOE of primary energy needs to be supplied using hydroelectricity. There is a convention to use 38% as the efficiency of thermal power plants to convert primary energy to the corresponding hydroelectricity. Hence, to save 1440 MTOE of primary energy from fossil fuels, there needs to be a generation of 547 MTOE worth of hydroelectricity. Converting to metric units, there is a need to generate about 23 EJ worth of hydroelectricity per year. Assuming a capacity factor of 60%, the corresponding hydroelectric power generating capacity that needs to be installed is about 1200 GW. Compare these numbers with the present installed capacity of 675 GW and a hydroelectricity generation of 2.3 trillion kWh, or about 8 EJ [ICOLD 2012]. It took more than a century to establish this capacity while a much larger capacity addition is necessary within a short time period to replace just a small part of the fossil fuel usage. Quite clearly, while hydroelectricity is a very important resource, it won't be able to replace fossil fuels.

Summary of Energy Benchmarks for Hydroelectricity:

Table 5.2 summarizes the energy benchmarks for hydroelectricity. While hydroelectricity is a wonderful renewable energy resource due to its dispatchable nature, low cost, and high EROEI, it is characterized by low power density, which leads to a high land-use. Hydroelectricity has a tremendous potential, especially in the developing nations in Asia and Africa. These nations can increase the per capita availability of electricity by harnessing the hydro potential. However, it is necessary to keep in mind that large dams do cause ecological damage, and that will be the price to pay for the economic development made possible by a higher availability of electricity.

Capacity Factor	Up to 100%
Dispatchability	Fully Dispatchable
Peak Power Density	2 to 25 W/m^2
Energy Density	3 to 80 $kWh/m^2/year$
EUF	About 0.5
EROEI	About 200
Cost	Very low

Table 5.2: Energy Benchmarks for Hydroelectricity

Pumped Storage Systems

A pumped storage system is a large, man-made reservoir, situated at a height. Water is pumped into this reservoir with high capacity pumps using power from the electric grid. These pumps are operated only during the interval when excess power is available on the grid. Water stored in the reservoir is used to drive turbines to generate electricity during the intervals of peak load on the grid to meet the excess power demand. In effect, the pumped storage system functions like a very large battery that is charged and discharged as needed. Building such a reservoir is feasible and cost effective only in locations where geography and geological characteristics are suitable. According to the National Hydro Power Corporation Ltd. (NHPC) of India, the potential for pumped storage systems in India is about 94,000 MW. Such systems work well with solar and wind energy systems which provide power intermittently. During the intervals when power is available, water is pumped up into the reservoir and electricity is dispatched using the pumped storage systems based on demands from the electric grid.

Wind Energy

It is a well known fact that the wind contains energy. Ships and windmills were the earliest devices to make use of this energy. In the modern times, a wind turbine is used to convert the kinetic energy of air to electrical energy. Conversion into electrical energy makes it possible to decouple the generation with usage, as electricity can be carried over long distances. A tremendous boom in the wind turbine installations was seen in the last ten years (2001 to 2011). At the end of the year 2011, the cumulative wind energy capacity installed in the world stood at 241 GW [Ryan and Bollinger 2012]. Many nations have programs to aggressively harness wind energy in the coming years. It will be be interesting to understand and evaluate the characteristics of this renewable energy source to compare with other sources of energy.

A wind turbine consists of a tower that rises 50 to 100 meters above the ground. Turbine blades are connected to a shaft mounted on the tower. Turbine blades are 30 to 50 meters long and sweep a large area as they rotate due to wind, converting the kinetic energy of air to electricity with the help of a generator. Wind turbines also contain safety mechanisms for protection against very high wind speeds. Depending on the design, wind turbines may contain gearbox, electronics, sensors for wind direction, and a yaw motor to position the rotating blades in the path of the wind.

Wind Speed (m/s)	Power W/m^2
1	0.6
2	5
5	77
8	315
10	615
12	1063

Table 5.3: Wind Speed and Power Density

The energy contained in the wind is represented by wind speed which is measured in meters per second (m/s). The wind speed can range from very gentle wind speeds of less than 1 m/s to devastating hurricanes with wind speeds of more than 70 m/s. The

Illustration 5.2: Offshore Wind Turbine in Northumberland, United Kingdom. Credits: Crepesoles/Shutterstock.com

power and energy content of wind is proportional to the cube of wind speed. Thus, a category 5 hurricane with a wind speed of 70 m/s is 40 times powerful as compared to a tropical storm with a wind speed of 20 m/s and 343,000 times powerful as compared to a gentle breeze at an air speed of 1 m/s. Table 5.3 lists the wind speeds and the corresponding power densities on a vertical surface mounted perpendicular to the direction of wind. There are many geographic areas in the world where average wind speed is more than 5 m/s at a height of 50 to 80 meters above the ground. Locations with an average wind speed of more than 10 m/s are comparatively fewer. Hence, a number of locations in the world would have average wind power densities of 75 to 600 W/m². Generally, a power density greater than 250 W/m² is considered suitable to harness wind energy. Note that this is the power density on a vertical surface. The calculation for power density on a horizontal surface, i.e. in a geographic area, will be covered later in this chapter. Only a part of the energy incident on the vertical surface gets converted into electrical energy. The theoretical maximum conversion efficiency is 59.3%, known as the Betz limit. A practical wind turbine does not achieve more than 40% efficiency. For instance, the Suzlon S97 wind turbine has a maximum power specification of 2.1 MW for a rated wind speed of 11 m/s, which corresponds to a power density of 818 W/m^2 [Suzlon 2011]. This turbine has a rotor diameter of 97 meters which implies a swept area of 7,386 m^2. The power input for this area works out to be about 6 MW, which implies an output efficiency of about 35%.

Wind turbines can also be installed on offshore locations. There is no obstruction to the wind flow on the seas and such turbines yield a better output. Offshore turbines are generally located close to the coastline to limit the installation costs. Illustration 5.2 shows an offshore wind turbine located in the United Kingdom.

Wind Energy: Worldwide Usage

Wind energy usage has increased all over the world at a fast pace in the last 10 years. India, USA, and China added, 3,300 MW, 6,816 MW, and 17,631 MW respectively, of new capacity in the year 2011. Table 5.4 shows the cumulative wind power capacities at the end of the year 2011 for the top 5 nations [Wiser and Bolinger 2012].

Country	Cumulative Wind Power Capacity (MW)
China	62,412
USA	46,916
Germany	29,248
Spain	21,350
India	16,266

Table 5.4: Cumulative Wind Power Capacity in the Year 2011

Nations in the Western Europe were the early adapters of wind energy and these nations lead in terms of the percentage of total electricity consumption supplied by wind energy. Denmark was the world leader in 2011, as wind energy supplied about 29% of the total electricity consumption. Wind energy supplied more than 10% of the total electric consumption in Portugal, Spain, Ireland, and Germany. China, USA, and India lag in this number though these countries lead in the new installations. It will be useful to determine the percentage of the Total Primary Energy (TPE) consumption that wind energy supports for the top five nations listed in Table 5.4. To calculate this, some assumptions need to be made:

- It will be assumed that the wind power capacity at the end of 2011 was available for the entire year.
- Capacity factors for wind energy tend to be different for every country. They will be covered in detail in a subsequent section. For the purpose of this analysis, it is sufficient to use the following numbers: USA 25.7%, Germany 18.3%, Spain 24.8%, India 21%, and China 21%. The data for Germany, Spain, and USA is based on a paper by Nicolas Boccard [Boccard 2009] while the data for India is based on a report by the Global Wind Energy Council [GWEC 2011]. In absence of a very reliable source, the capacity factor for China is assumed to be the same as India.
- To convert wind electricity to heat units, it will be assumed that the conversion efficiency of thermal power plants is 38%. This will allow the calculation of primary energy saved as a result of wind energy usage.

Country	Wind Energy (MWh)	Primary Energy for Wind (EJ)	TPE (EJ)	Wind Energy (%)
USA	105,622,929	1.001	95.3	1.05
China	114,813,115	1.088	109.8	0.99
India	29,922,934	0.283	23.5	1.21
Germany	46,886,884	0.444	12.9	3.45
Spain	46,382,448	0.439	6.1	7.17

Table 5.5: Wind Energy as a Percentage of TPE Consumption in the Year 2011

It can be observed from Table 5.5 that wind energy supports only a small percentage of the TPE consumption in spite of the rapid pace of new installations over the last decade. What is the potential that wind energy has to grow from these dismal levels and what are the obstacles to this growth? Next few sections attempt to address these questions.

Capacity Factor for Wind Energy

The wind speed not only changes throughout the day but also varies seasonally within a year. In fact, there are inter-year variations too. As a result of the variations in wind speed, electrical output from a wind turbine changes all the time. A wind turbine is designed to provide the maximum power output at a certain wind speed, known as the rated speed. At very low wind speeds, turbine blades may not even rotate and there will be no electrical output. At other wind speeds, the output will be equal or lower than the rated maximum output. Thus, a turbine rated at 1 MW will not produce 24 MWh of electrical output in one day. In most locations suited for wind turbine installation, the output will range from 5 MWh to 10 MWh, on the average. This implies a capacity factor that ranges from 20% to 40%. Wind turbines can also be installed offshore and the capacity factor tends to be higher for the offshore installations. Table 5.6 shows the variability of wind energy generation based on data for all wind turbine installations in Maharashtra, India [MEDA 2009]. The generation is at its peak during the month of July in the monsoon season which lasts from June to September. Generation falls rapidly after the withdrawal of monsoon and is at its lowest in January and February. Wind speeds increase before the onset of monsoon and a good amount of generation takes place in the month of May. Generation in the month of January is just 18% of the peak generation in July.

Month	Generation (GWh)	Month	Generation (GWh)
2008 April	167	October	100
May	397	November	115
June	391	December	94
July	441	2009 January	78
August	415	February	82
September	212	March	137

Table 5.6: Variability in Wind Power Generation, Maharashtra 2008-09

Capacity factor is important for two reasons:
- Capacity factor directly translates into energy delivered and hence the return on investment. It is desirable to have a high capacity factor for a good return on investment.
- One of the reasons for installing wind turbines is to meet certain emissions targets. A higher capacity factor implies a higher generation of wind energy.

The corresponding savings in fossil fuels directly correspond to savings in CO_2 emissions.

Achievable capacity factor is an important criteria while selecting suitable sites for generating wind energy. Certain locations are known to be windy and capacity factors tend to be high in such locations. The plains of central USA that range from Texas to North Dakota and the coastline of Spain and Ireland are examples of such locations.

Wind Energy: Resource Potential

Kinetic energy is imparted to air as a result of the direct solar insolation on earth. The total kinetic energy of air is probably several orders of magnitude more than the energy usage by mankind. However, only a small fraction of this wind energy is accessible via the lowermost layer of the atmosphere. The towers used for tapping wind energy have a height that is generally less than 100 meters. Yet, the total accessible wind energy potential is very large.

Wind energy potential depends on the height at which the wind speed is measured. Potential increases with the height from the ground. Wind energy potential is normally specified at a height of 50 to 100 meters. For instance, the potential in India is estimated to be 49,130 MW at a height of 50 meters and 102,788 MW at a height of 80 meters [CWET 2012]. The potential also depends on the capacity factor achievable at a particular site. Higher the requirement of the capacity factor, lower the potential. National Renewable Energy Laboratory (NREL), USA, provides state-wise estimates of wind energy potential, at various capacity factors. Most of the potential is available at capacity factors less than 35%. The potential is almost 0 above 50% capacity factor. These capacity factors are theoretical and the realized capacity factors can be less.

The potential for wind energy in the US is huge at a capacity factor of 30% or better. Just the two states of Kansas and Iowa, together have a potential to generate 1,523 GW with an annual generation of 5.6 million GWh of electricity [NREL 2011]. Compare these number with a total installed capacity of 1,039 GW for all sources of electrical energy in USA with an annual generation of 4.1 million GWh [EIA 2011]. In other words, harnessing the total wind energy potential in Iowa and Kansas will fulfill the entire electricity demand in the US, at least in theory! Even in India, the estimated potential of 102 GW is probably an underestimate since it assumes a land availability of just 2% in windy regions. As energy prices increase and ecological hazards of coal mining become increasingly unacceptable, land availability for wind energy should increase. Secondly, the offshore potential is not accounted for in these calculations and is expected to be very large with a better capacity factor.

In summary, the potential is certainly not a problem in the large scale use of wind energy. Wind energy has a much lesser impact on environment as compared to the large dams required for hydroelectric energy. Seismic sensitivity of the region does not pose any risk either. Yet, the progress of wind energy has been slow, as seen in Table 5.5. There are a number of fundamental reasons for this slow progress, which will be explored in the following sections.

Further Reading on the Web:

The National Renewable Energy Laboratory (NREL) of the US Department of Energy provides some very useful information about wind energy technology and resources on its website:

NREL: http://www.nrel.gov/wind/

Power Density and Energy Density for Wind Energy

It is fairly easy to calculate the power density and energy density for wind energy. The power density on the vertical plane is dependent on wind speed. The power density in the horizontal plane is dependent upon following factors:

- Power density in the vertical plane
- Distance between turbines
- Turbine efficiency

It is easy to derive the equation for power density. Let, w be the power density in the vertical plane, r the radius of the turbine rotor, s the number of diameters of spacing between turbines, and η be the turbine efficiency. In this case,

Power incident on the area swept by rotor = $\pi * r^2 * w$

Electrical output from turbine = $\eta * \pi * r^2 * w$

Area on Ground for Each Turbine = $(s*2*r)^2 = 4s^2r^2$

Power Density = Output/Area on Ground = $(\eta \pi r^2 w)/(4s^2r^2) = (\eta \pi w)/(4s^2)$

It is reasonable to assume the turbine efficiency to be 40%, spacing between turbines to be 10 rotor diameters, and power density in the vertical plane to be 400 W/m^2. Using these numbers, power density for wind energy is calculated to be about 1.3 W/m^2. It can be observed that the power density depends on the spacing between turbines. The assumption of 10 rotor diameters is somewhere between the conventional number of 7 and a recent study which suggests that 15 diameters is the optimal spacing [Johns Hopkins University 2011]. At a capacity factor of 25%, the power density of 1.3 W/m^2 translates into an energy density of 2.8 kWh/m^2/year. The numbers for power and energy density for wind energy are lower than those for hydroelectricity. However, this low power and energy density is not a major problem as far as land-use is concerned. It is possible to use the land between wind turbines for farming and grazing. This is different from hydroelectricity, where a huge area is immersed in the reservoir of a dam.

The real problem with the low energy density is the infrastructure that will be required to build and operate wind farms over a very large geographic area. At a power density of 1.3 W/m^2, harnessing 50 GW from wind energy will require wind farms spread over an area of 38,400 square kilometers! Wind turbines are very large in size. Their construction and subsequent maintenance requires movement of heavy equipment for which suitable access roads must be built. In many cases, roads need to be built in difficult terrain, which is quite expensive. A low energy density also implies that the electric grid needs to be extended to reach the arrays of turbines spread out over a large area. Such an extension of the electric grid is an expensive proposition, considering the rising price of copper. In many cases, these infrastructure activities are carried out in the form of a subsidy by the government to meet certain social objectives. Such a policy would work initially when sites close to roads and the power grid are tapped. It remains to be seen if such policies will scale when wind farms need to be built in remote areas.

The Problem of Predictability and Dispatchability

Wind energy is not a dispatchable source of energy. There is absolutely no control over when and how much power will be generated. The power output solely depends on the wind speed. It is very difficult to make a precise prediction about the expected wind speed at a given time of the day at each of the locations where wind turbines are installed. As a result, it is impossible to predict the exact output of wind turbines at a specific time. There can be fairly accurate predictions about the average output in a given season. However, it is not possible to predict the exact output on a daily, hourly, or on a minute by minute basis. This lack of predictability and the variation in the output makes the job of the electric grid operator very difficult. While it is obvious that wind energy cannot serve the base-load on the grid due to the low capacity factor, serving the peak loads is not easy due to the lack of predictability and variability. Secondly, the peak in the output of wind turbines may not correspond to the time of peak load on the grid. Connecting a large number of wind turbines to the grid from different geographic regions in a country does help as there is some predictability to the minimum output that will be available. Yet, it is necessary to complement wind energy with hydroelectricity or natural gas based thermal power stations, which can generate power on short notice when wind energy is not available. Such a requirement drives up the total cost and also limits the amount of power that can be sourced from wind energy. It will be a challenge to source more than 20% of the electrical energy from wind in most countries where capacity factor tends to be less than 30%.

Cost, EUF, and EROEI for Wind Energy

Wind power is considered to be an inexpensive source of energy. In USA, the installed cost of wind turbines is about $2,000 per kW [Wiser and Bolinger 2012]. It is possible to estimate the EROEI and EUF for wind farms in USA based on the installed cost and the energy intensity of US economy, with certain assumptions. The US economy is primarily a service sector economy. The service sector has a low energy intensity as compared to the industrial activity required for the production of wind turbines. It is assumed that the business of making wind turbines has an energy intensity that is thrice that of the US economy. Admittedly, this assumption will yield only an approximation of the EUF and EROEI. However, for the purpose of this book, that is what is the intention. The calculation of EUF and EROEI is presented in a tabular form in Table 5.7. It can be seen that the EUF is calculated to be about 2 with an EROEI of 12. A similar calculation for wind farms in India yields an EUF of 2.3 with an EROEI of 10.7. Wind turbines in India generally operate at a lower capacity factor as compared to the US. To calculate the best case EUF and EROEI for wind farms in the US, some assumptions will need to be changed. Assuming a 30% capacity factor along with an energy intensity of the wind turbine business to be twice that of the US economy, EUF and EROEI are calculated to be 1.1 and 21.7, respectively. It should be noted that these calculations do not include the cost of building roads and major upgrades to the electric grid. As wind farms are built in remote and sparsely populated locations, these factors must be taken into consideration. Secondly, these calculations are not for offshore wind farms which are expected to be quite expensive while yielding a better capacity factor. Another important factor is the non-dispatchable nature of wind energy which necessitates a backup power source on the grid. The cost and the corresponding energy input required for such sources is not included in the calculation.

Parameter	Value	Comments
Installed Cost Per Watt ($)	2	Ref: [Wiser and Bolinger 2012]
Cost Inflation adjusted to 2009 Dollars	1.91	
Energy Intensity of US Economy (MJ/$)	7.52	
Adjusted Energy Intensity for Wind Turbines (MJ/$)	22.56	Thrice the US energy intensity
Energy Input per Watt (MJ)	43.06	
Energy Output per Year per Watt (Wh)	2190	Assumes 25% capacity factor
Energy Output Per year (MJ)	7.88	
Primary Energy Saved per Year (MJ)	20.75	Efficiency of thermal power plants = 38%
EUF	2.08	
EROEI	12.04	Lifespan of 25 years with negligible operational costs.

Table 5.7: EUF and EROEI Calculation for Wind Energy

Summary of Energy Benchmarks for Wind Energy

Table 5.8 lists the energy benchmarks for wind energy. In summary, wind energy has a low power/energy density, low EROEI, and a higher EUF, as compared to hydroelectricity. A higher EUF implies that a significant amount of energy must be consumed upfront. This is especially true for wind farms in remote locations and difficult terrains that require a major effort to build the infrastructure. In the case of hydroelectric power, the actual construction of dam takes place on a relatively small area as compared to the size of the reservoir that is created. Thus, with a relatively small effort, a highly concentrated source of energy is created. However, in the case of wind energy, the creation of a concentrated source of energy takes place only as a result of extending grid connectivity to each and every wind turbine, spread out over a large area.

The potential for wind energy is huge but the non dispatchable nature and the low capacity factor makes it a less dependable resource. As a result, the penetration of wind energy may not increase much beyond 20% of the total electrical energy delivered.

On the positive side, wind energy is a clean source of energy. The CO_2 emissions are limited to the manufacturing and installation of wind turbines. The ecological problems of wind energy are very few and the widespread potential makes it a democratically accessible resource. The technology to harness wind energy is well developed and efficient. There should be a rapid growth in the usage of wind energy, subject to limits and limitations mentioned earlier.

Capacity Factor	18 to 35%
Dispatchability	Not Dispatchable
Peak Power Density	About 1.3 W/m²
Energy Density	About 2.8 kWh/m²/year
EUF	About 2
EROEI	About 12
Cost	Medium

Table 5.8: Summary of Energy Benchmarks for Wind Energy

Solar Photovoltaic Electricity

A solar cell is a device that directly converts sunlight into electricity. The beauty of this device is the direct conversion of sunlight into electrical energy, with no intermediate steps. It is a clean and operationally simple method of energy conversion. There are no moving parts and a long operational life with very little maintenance is guaranteed.

A solar cell is not a new device. It is based on the photovoltaic effect: Certain materials, when exposed to sunlight, produce an electromotive force (EMF) that has the ability to drive a current in an electrical circuit. This effect was discovered by Edmond Becquerel, a French Physicist, in 1839. However, it was only in the 1950s that Bell Labs developed a silicon solar cell, similar to what is in use today. In the meantime, a number of materials, notably selenium, were discovered which demonstrated the photovoltaic effect. However, it was only the silicon solar cell that

Illustration 5.3: Solar Panels on the roof. Credits: Smileus/Shutterstock.com

converted a reasonable amount of sunlight into energy. Since its discovery in the 1950s, a number of improvements have been made in the silicon solar cell to convert more sunlight into electrical energy.

Output voltage from a single solar cell is only about 0.6 Volts, while the output current depends on the area of the solar cell and the sunlight incident on the solar cell. Output voltage of a single solar cell is too low to be useful in an electrical circuit. To solve this problem, a number of solar cells are connected in series in a solar panel. The voltage output of a solar panel is the sum of voltage outputs of the individual solar cells. Since the solar cells are connected in series, an identical current passes through each of them. As a result, it is necessary that all the solar cells in a panel are matched to produce an identical current. In the case of any mismatch, the solar cell producing the least current will dictate the current output from a solar panel.

A solar panel has a glass shield and is sealed for protection from harsh environment. It also has a mounting arrangement and electrical connectors. Solar panels have a very long life with no maintenance except cleaning. Most commercially available solar panels provide a 20 to 25 year warranty.

Using Solar Panels

In Silicon Valley, California, some single-family homes have solar panels installed on the roof. These panels use a synchronous inverter to generate a 110 Volts AC supply that is in perfect synchronization with the power received from the electric grid. The output from the solar panel is first utilized to satisfy any household demand. Any excess power that is available is fed back into the grid. When the grid is supplied with power from a rooftop solar panel, the electric meter actually runs in the reverse direction. In fact, there are smart electric meters that price electricity based on the time of the day. Electricity delivered during the hours of peak demand gets a premium rate. Attractive rates from utility companies along with subsidies from the government has made it possible for some homeowners to invest the capital required for a rooftop system. There are several regions in the world, primarily in the affluent OECD nations, where such applications of solar energy are growing, thanks to government subsidies. The motivation partly comes from targets to reduce the emissions of greenhouse gases. Secondly, it is a reasonable argument that subsidies create a demand that will result in economies of scale. A higher demand also attracts investment and R&D activity, which can help reduce the price of solar panels.

Apart from the household rooftop installations (see Illustration 5.3), solar panels are also used in large numbers in solar farms to produce electricity on a large scale. Such solar farms are typically located in sunny locations where land is available at a reasonable price. There are industrial and commercial uses of solar panels too, many of them using rooftop installations.

In India, it is still quite rare to see rooftop installation of solar panels on individual homes. Capital costs are very high and the solar panels don't provide an attractive return on investment with an insufficient subsidy from the government. The more common applications of solar photovoltaic electricity in India are:

- Street lighting in remote places
- Lighting up homes in remote villages
- Solar lanterns and emergency lighting
- Water pumps for agriculture
- Solar photovoltaic farms connected to grid

Yet, the situation is certain to change with the Jawaharlal Nehru National Solar Mission, that plans an installation of 20,000 MW of solar power by year the 2022 [Solar India].

The usage of solar photovoltaic energy lags behind wind energy by almost two orders of magnitude though the resource potential is significantly higher. For instance, in the year 2010, wind energy delivered 95 billion kWh of electricity in USA, while solar photovoltaic electricity accounted for only 1.3 billion kWh. The corresponding figures in India were, 20 billion kWh and 0.1 billion kWh, respectively [EIA]. The following sections will explore the technical details to understand the reasons for slow adaptation and to project the future direction for the solar photovoltaic energy.

Solar Insolation

It is known that locations close to the equator are considerably hotter than the locations close to the arctic circle. Even on the same latitude, two locations may have a difference in the perception of heat. Solar insolation is the amount of solar energy incident at a given location. It is measured in terms of Watts/m^2. Solar insolation is about 1000 W/m^2 when the sun is directly overhead at an extremely dry location such as the Sahara Desert. Effectiveness of a solar energy solution for a given location depends on the average amount of solar insolation at that location. The solar insolation changes throughout the day as well as throughout the year, as seasons change. There are inter-year variations as well. It also depends on factors such as the moisture content in air, cloud cover, and even the particulate matter or pollution in the air. How does one find out the insolation at a give location? Apart from the obvious choice of measuring it with suitable instrumentation, there are organizations that publish this data on a global scale. NASA carried out satellite observations of earth over 22 years to record some very useful data on solar insolation. Another alternative is the NREL database, available via web at: **http://www.nrel.gov/**

Insolation data is available in a very useful, graphical form on the NREL website. In general, the insolation data available via these organizations is correct at a macro level, and the resolution may not be better than 10 km^2. The data is collected by satellites orbiting around the earth that measure various atmospheric parameters. This data is processed to estimate insolation on, horizontal surface, tilted surface, and for a surface that tracks the sun. While estimating the feasibility of solar photovoltaic installation for a geographic region, this data is extremely useful. However, for a particular site, it necessary to take into account local factors such as, shadowing and micro-climate. The data collected using satellites needs to be supplemented by observations at the ground level.

Illustration 5.4 is an example of insolation data for India in the month of May. This map was published with permission from NREL. It can be observed that a large part of

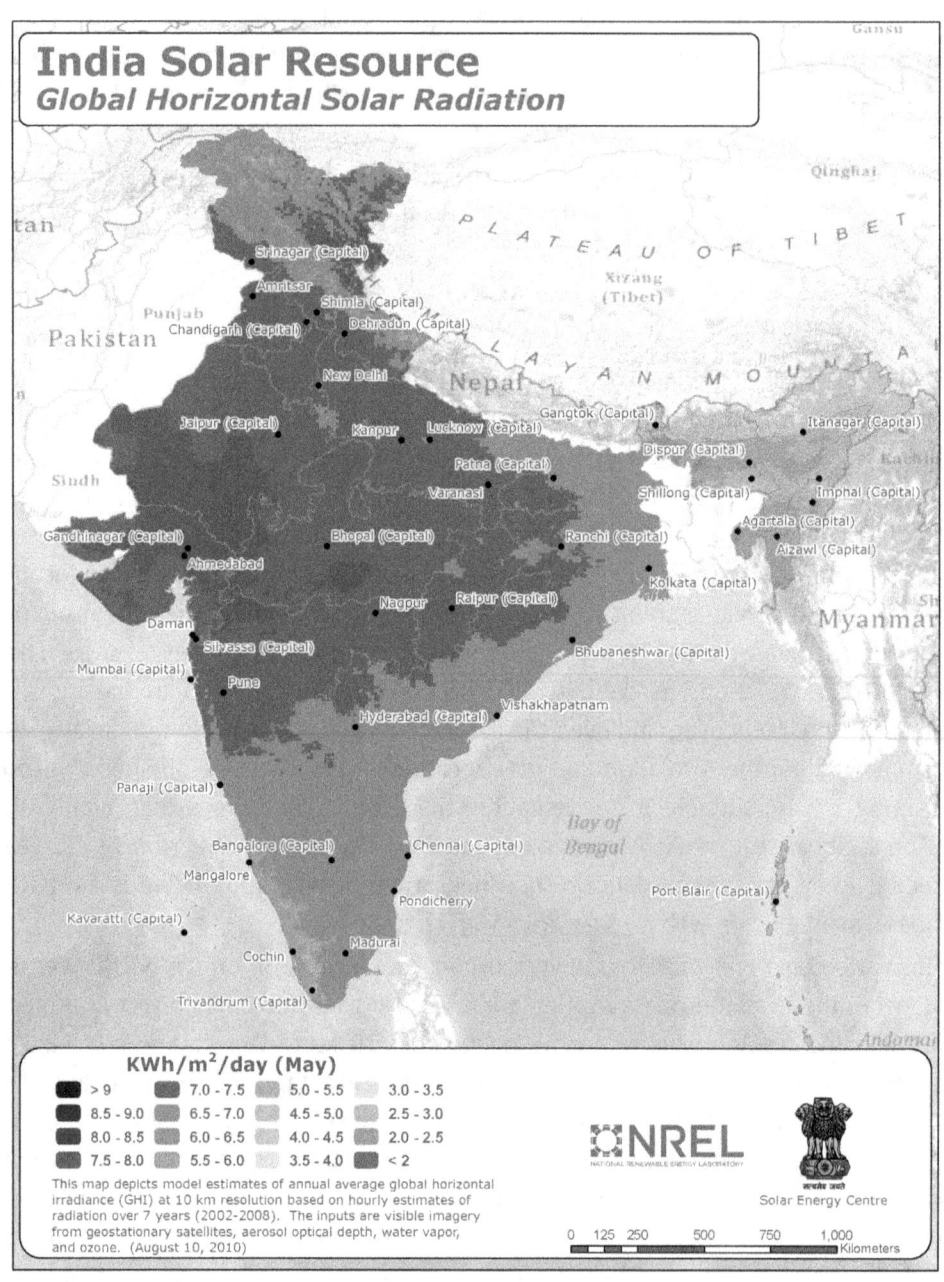

Illustration 5.4: Solar Insolation in India (May). This map was created by the National Renewable Energy Laboratory for the Department of Energy, USA

India receives about 7.0 to 7.5 kWh/m^2/day, with some parts receiving an even higher insolation. A simple calculation will illustrate the enormity of the solar insolation. Assuming an average insolation in the month of May in India of 7.0 kWh/m^2/day, with an approximate area of 3.2 million sq. km, the total Insolation for India is calculated to be 22.4 trillion kWh/day. In other words, a single day in the month of May corresponds to 80 EJ worth of solar insolation in India. Compare this with a TPE consumption of just 22 EJ in India in the year 2010. Quite clearly, solar insolation is an enormous resource and direct conversion using photovoltaic panels is an elegant method of conversion that needs an in-depth analysis. Next several sections are devoted to the analysis of photovoltaic electricity, followed by solar thermal electric conversion.

Solar Panel Specifications

Table 5.9 lists the parameters that specify the output of a solar panel.

Parameter	Description
P_{max}	Maximum power output from the panel, specified in Watts
V_{oc}	Open-circuit voltage, specified in Volts. This is the voltage output from the panel when no current is drawn.
I_{sc}	Current output from the panel in Amperes when the output terminals of the panel are shorted, i.e. connected to each other.
V_{mpp}	Voltage output from the panel when maximum power is being delivered.
I_{mpp}	Current output from the panel when maximum power is being delivered.

Table 5.9: Solar Panel Parameters

Note that V_{mpp} and I_{mpp} are always less than V_{oc} and I_{sc}, respectively. These parameters are specified at Standard Test Conditions (STC), which are the ideal conditions under which the output of a solar panel is specified. STC corresponds to a solar insolation of 1000 W/m^2 and a cell temperature 25° Celsius. The variation of these parameters with temperature and insolation is specified as well. Some of the specifications for the Kyocera KD-215GX solar panel are listed in Table 5.10 [Kyocera].

Parameter	Value
P_{max}	215 W
V_{oc}	33.2 V
I_{sc}	8.78 A
V_{mpp}	26.6 V
I_{mpp}	8.09 A

Table 5.10: Kyocera KD-215GX Parameter Specifications (partial)

Peak Power Point for a Solar Panel

Illustration 5.5 shows the voltage-current characteristics of a typical solar panel. Current is plotted on the X-axis and the voltage on the Y-axis. When a solar panel is not connected to any load, current output is zero while voltage output is maximum (V_{oc}). At this point, the power output from the panel is zero. As current drawn from the panel is increased, voltage remains almost constant, falling only slightly. As a result, the power output from the panel rises along with the current, until the *knee point* is reached. After this point of maximum power, voltage output from the panel rapidly falls with an increase in the current, resulting in a reduced power output. For obtaining

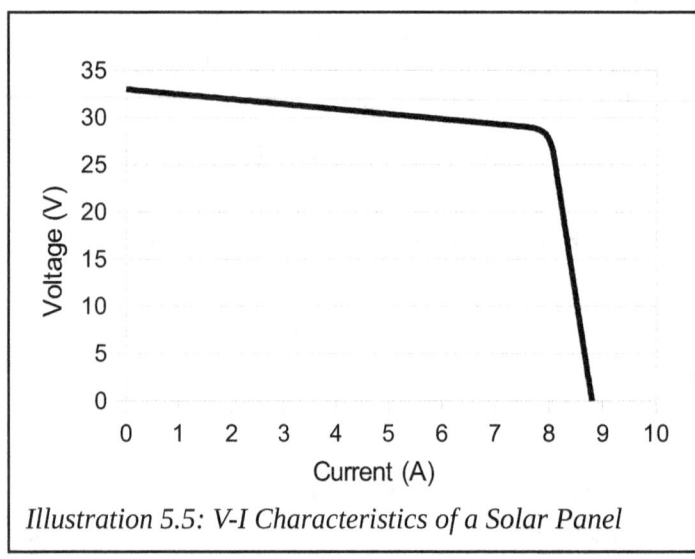

Illustration 5.5: V-I Characteristics of a Solar Panel

maximum energy output form a panel, it is necessary to draw a current that corresponds to the knee point. However, the V-I characteristics of a panel change with solar insolation, which undergoes a change throughout the day. As the insolation changes, the maximum power point changes too. Hence, continuous operation of a panel at the maximum power point requires some smart electronics.

Solar Panels and the Tilted Installations

Sun traverses a path in the sky that depends on the latitude of the location. On the equinox day, when sun rises exactly in the east, it comes directly overhead at noon at places located on the equator. At other locations in the northern hemisphere, sun appears tilted south at an angle equal to the latitude of the location. For instance, in Pune, India, on March 21st, sun will appear about 18° to south at noon, as compared to the point directly overhead. In San Francisco, USA, this angle will be about 38°. If a solar panel is placed horizontally, sun rays will not be incident perpendicular to the panel, even at noon. Output from a solar panel is maximum when the sun rays are incident perpendicular to the panel. By tilting the panel towards south at an angle that equals the latitude of the location, it is guaranteed that the sun rays will be incident perpendicular at noon on the equinox day. As the sun traverses the sky during the day, sun rays don't remain perpendicular to the panel. Even during the year, as the seasons change, sun does not rise exactly in the east. During summer, in the northern hemisphere, sun rises towards the north-east direction, while in the winter, sunrise takes place in the south east direction. This too results in the sunlight not being incident perpendicular to the panel, even at noon. However, on the average, tilting the panel at latitude degrees towards south minimizes the angle of incidence, resulting in maximum output from the solar panel. This concept applies not only to photovoltaic panels but also for applications such as solar water heaters.

Sometimes, it is necessary to maximize the power output during the winter months. For solar water heating applications, for many places in India, it is necessary to optimize heating during the winter months. On the day of winter solstice, at noon, sunlight is incident at an angle of latitude + 23.5° to the perpendicular, for a horizontally placed panel. Thus, for Pune, India, the angle of incidence would be about 42° at noon on December 22nd, which is the day of winter solstice. This is the maximum angle that will be subtended. The angle of incidence on the day of equinox is about 18°. To optimize the output during the winter months, panel is tilted at about 30° angle towards south, which is the midpoint of the two angular positions. Quite obviously, optimizing the output during winter months implies that slightly lower output will be available during the summer months. For applications such as water heating, this is not a problem for most locations in India. Summers are too hot to require any water heating!

As a corollary, there will be other locations that will require optimization during the summer. Seattle, USA, is probably a candidate for this. In Seattle, during the winter months, sky is overcast on many days and more diffuse solar radiation reaches the ground than direct sunlight. However, during the summer, Seattle enjoys clear skies and long daylight hours with ample solar energy reaching the ground. For Seattle, it will be correct to optimize the solar energy conversion during summer months by tilting the panel towards south at an angle that is less than the latitude for Seattle.

In general, the idea is to tilt the panel at an angle that will maximize the output with respect to the local requirements. Average annual insolation for a few selected places in India and USA is presented in Table 5.11. A solar photovoltaic panel or even solar water heaters are expensive resources and maximizing their outputs is a very important requirement.

City	Average Annual Insolation on Horizontal Surface (kWh/m²/day)	Average Annual Insolation on Surface Tilted at Latitude Degrees (kWh/m²/day)
Bakersfield, CA	5.31	5.80
Bangalore, India	5.25	5.37
Boston, MA	3.70	4.20
New Delhi, India	5.05	5.48
Pune, India	5.11	5.41
San Jose, CA	5.30	5.85
Seattle, WA	3.36	3.60

Table 5.11: Average Annual Insolation, USA and India, Selected Cities (Source: Analysis with the help of [NASA EOS])

Efficiency of a Solar Photovoltaic Panel

Efficiency of a solar panel is typically about 14%, which implies that around 14% of the solar energy incident on the panel is actually converted into electricity. In Pune, India, average solar insolation on a horizontal panel is about 5.11 kWh/m²/day. In ideal conditions, a solar panel of size 1 square meter with a 14% efficiency will output about 0.72 kWh of electrical energy in a day. The efficiency depends on the type of solar cells used and how densely the cells are packed. Efficiency of single-crystal silicon cells can be more that 20%, while the efficiency is low for amorphous silicon cells. Polycrystalline solar cells have a slightly lower efficiency as compared to the single-crystal cells but cost less. There are multi-junction solar cells that have reached efficiencies of 40%. However, these solar cells are extremely expensive. They are either used in space applications or along with tracking and concentration devices, which are covered later in this chapter. Polycrystalline solar panels are generally the most cost effective.

Capacity Factor for Solar Photovoltaic Power

Output of a solar panel is specified in Watts for Standard Test Conditions (STC). The STC requires a solar insolation of 1000 W/m² and a cell temperature of 25° Celsius. For instance, the Kyocera KD215GX-LPU panel is rated to produce 215W as the maximum power when solar insolation is 1000 W/m² and solar-cell temperature is limited to 25° Celsius. This output power is rarely achieved in practice. There are several reasons for the reduction in output. Each of these reasons correspond to a down-rating factor to apply to the rated power. In the following paragraphs, down-rating factors due to various other causes are calculated.:

- Solar insolation: An insolation of 1000 W/m² is probably achieved in the Sahara desert when the sun is directly overhead! In practice, the insolation varies considerably across the globe. Absorption of sunlight in air varies, depending on the volume of air through which the sunlight passes. The volume depends on the latitude and the season. At latitudes far away from the equator, sun never comes directly overhead and the sunlight must pass through a larger volume of air. During the winter season, there is a further increase in the volume of air through which the sunlight must pass, as the declination of sun in the sky increases. Absorption in the atmosphere also depends upon factors such as humidity, dust particles, and aerosols present in the air. Moreover, the insolation reduces significantly in cloudy, overcast conditions. Given these variables, how do we estimate the amount of energy that a solar panel will deliver in a year? Thankfully, NASA has gathered data for solar insolation for almost every part of the globe and averaged it over the past 22 years. This data was gathered using satellites. Analysis of this data reveals that for many cities in India, the average insolation is only a little over 700 W/m² for a panel that is tilted at an angle of latitude degrees towards south. For instance, for the city of Pune, the average solar insolation is calculated to be 708 W/m² for a panel that is tilted at an angle of 18 degrees towards south. The main reason for this deviation, as compared to the test condition of 1000 W/m², is the impact of the monsoon season. During the months of July and August, the amount of sunlight reaching the ground is just over half of that in the months of March and April, in spite of more daylight hours! Hence, a calculation for energy output needs to down-rate the solar panel wattage based on the insolation data for the location of installation. Average insolation for multiple cities in India is calculated to be 717 W/m². Thus, we arrive at a down-rating multiplicand of 0.717 to down-rate the power output from the solar panel.

- Operating Temperature: A solar panel rarely operates at a cell temperature of 25° Celsius. In a tropical country like India, it is more common for the cell temperature to rise to more than 50° Celsius. At this temperature, the power output from a solar panel is less as compared to that at 25° Celsius. For the Kyocera panel, about 10% drop in power output is expected at the cell temperature of 50° Celsius [Kyocera]. Hence, the down-rating factor due to temperature of operation is calculated to be 0.9.

- Electronics associated with using the energy produced by the solar panel can't operate at 100% efficiency. Irrespective of whether the energy produced by the panel is used to feed the grid or to drive a domestic AC or DC load, the efficiency is generally about 85%. This includes the fact that it may not always be possible to operate the panel at the peak power point. Thus, down-rating factor due to efficiency of the associated electronics is 0.85.

- Solar panels collect a lot of dust on the glass surface. This reduces the output from the panel. Even when the dust is cleaned periodically, some residue remains. Reduction in output of even a single solar cell affects the power output from the entire panel! Remember, all the solar cells within the panel are connected in series, and the cell with the lowest current output dictates the current output from the solar panel. Likewise, any problems in manufacturing, or reduction in output current due to aging of a single solar cell, affects the current output from the solar panel. Reduction in the current output causes a corresponding reduction in power and energy output of the solar panel. The reduction in output of a solar panel due to these factors is mostly a guess! An aggressive estimate is a 5% reduction in the power output owing to the mismatches created due to dust, aging, etc. Thus, the down-rating factor on this account will be 0.95.

- COS(θ) Effect: During the course of the day, as sun traverses the sky, the angle of incidence of sunlight on the solar panel keeps changing. Early in the morning, and late in the evening, the angle of incidence is maximum. Near noon, this angle is 0 and maximum solar radiation in incident on the panel. The direct solar radiation incident on the panel early in the morning and late in the evening is close to 0. The radiation incident on the panel when the angle of incidence is, say 60 degrees will be half of that when the angle is 0 degrees. Thus, the 100W panel does not output 100W throughout the day, simply because sun is moving in the sky! Using some simple mathematics, the average radiation incident on the panel in a day is calculated to be about 0.64. This is called the $COS(\theta)$ effect. The performance degradation due to the $COS(\theta)$ effect is separate from the degradation caused by the lower average insolation. There is one more factor that results in another $COS(\theta)$ effect. Throughout the year, as seasons change, the declination of sun in the sky changes. This results in another angle of incidence, in a different plane. The $COS(\theta)$ effect of this is relatively little and is calculated to be 0.97. Thus, the down-rating factors due to the two $COS(\theta)$ effects will be, 0.64, and 0.97.

- Daylight Hours: The most obvious factor is the number of hours of daylight. On the average, sun shines for 12 hours in a day. This results in a down-rating factor of 0.5 for obtaining the average power available from the solar panel.

Multiplying all the down-rating factors, the net down-rating factor is calculated to be 0.16. Thus, the energy delivered by a 100W panel in a day is actually equivalent to the energy produced by a 16W power source throughout the day. In other words, it can be stated the capacity factor for solar photovoltaic power generation is about 16%, for an average location in India.

A similar calculation can be made for various locations in the US. Due to the vast deference in latitudes and weather patterns, capacity factors vary widely across the US. Calculated capacity factors for various cities in the US are listed in Table 5.12.

City	Capacity Factor	City	Capacity Factor
San Jose, CA	17%	Seattle, WA	11%
Orlando, FL	15%	Bakersfield, CA	17%
Boston, MA	12%	Phoenix, AZ	17%

Table 5.12: Photovoltaic Power: Capacity Factors for US Cities

How Much Energy does a 100W Solar Panel Generate?

With a capacity factor of 16%, it is easy to calculate the energy output in day. A 100W solar panel is equivalent to an energy source of 16W that operates 24 x 7, throughout the year. Thus, a 100W solar panel will deliver 384 Wh of energy in a day, and in one year it will deliver 140 kWh of energy. This calculation is an average for India, taking into account many major cities. Quite obviously, this number will vary, depending on the location. Secondly, even for a given location, there will be days when the energy output will be significantly larger or smaller. However, for most cities in India, the average energy delivered will be close to this number. The exceptions will be the locations in the far north-east and the northern-most parts of India, where solar insolation is much lower.

It will be interesting to calculate the corresponding numbers for some of the locations in the USA. For instance, at Seattle, WA, the expected energy output from a 100W panel is calculated to be only about 250 Wh/day, while at Bakersfield, CA, the same panel will output about 406 Wh/day. On the east coast, output at Orlando, FL, is expected to be about 345 Wh/day, while at Boston, MA, the output is expected to be only about 295 Wh/day. In Phoenix, AZ, the energy output is expected to be 413 Wh/day. While there is a wide variation in the energy output across the US, there exists a vast expanse of land in the eastern parts of Southern California, Arizona, Nevada, New Mexico, and Texas, where there is ample sunshine. Such locations are ideal for solar photovoltaic electricity generation.

Power Density and Energy Density for Photovoltaic Systems

Power and energy density are primarily important for large installations. Modern office buildings and shopping malls are characterized by a very high electricity consumption per square meter of floor space. A rooftop solar photovoltaic system may be suitable only if it generates sufficient amount of power to satisfy the consumption requirements. Likewise, land-use by commercial photovoltaic energy farms can be estimated on the basis of power and energy density calculations. Land-use can have a significant impact on the cost-effectiveness of a solar energy solution.

Power density calculation is based on the peak output power under standard test conditions (STC). The Kyocera KD215GX-LPU panel [Kyocera] has a peak power output of 215 Watts with a surface area of 1.48 m^2. The peak power density for this panel is calculated to be 145 W/m^2 when placed on a horizontal surface. There are other manufacturers which design panels with higher power density. For instance,

SunPower Corporation of San Jose, CA, makes solar panels that have 20% efficiency [SunPower 2011]. These panels provide a power density of 200 W/m^2 when mounted on a horizontal surface. However, panels are normally mounted with a tilt that equals the latitude of the location to maximize energy production. In such a case, shadow of one panel may obstruct the sunlight incident on another panel. To avoid this, it is necessary to maintain space between rows of panels which results in a reduction of power density. The farther the distance of a location from equator, the larger will be the tilt, requiring more space between rows of panels and hence a lower the power density. In general, power density will vary from 80 to 200 W/m^2, depending on the choice of solar panel and location.

The energy density calculation is also dependent on location. As mentioned earlier, a 100W solar panel delivers an average yearly energy output of 140 kWh at most locations in India. Thus, a power density of 130 W/m^2 corresponds to an energy density of about 180 kWh/m^2/year. This is the expected average across most of India. At Phoenix, AZ in USA, output from a 100W panel was calculated to be 413 Wh/day. For a peak power density of 130 W/m^2, the energy density works out to be 196 kWh/m^2/year. In the best case, for a power density of 200 W/m^2 at a location near the equator, energy density could be as high as 300 kWh/m^2/year, assuming a capacity factor of about 17%. Thus, at locations where capacity factor is 15% or more with power density of at least 80 to 200 W/m^2, energy density will range from, 100 to 300 kWh/m^2/year.

EUF and EROEI for Photovoltaic Systems

Opponents of solar energy have always questioned if solar photovoltaic systems deliver any net energy. This criticism has certainly decreased over the past decade as energy efficiencies improved and costs reduced. It is still worthwhile to make an attempt to calculate EUF and EROEI for solar energy. This calculation is done on the basis of some hard data and some assumptions. It needs to be stressed that EUF and EROEI calculation cannot be exact and does involve some guesswork. Calculation takes place in several steps:

1. **Price Determination:** Lawrence Berkley National Laboratory, USA, stated in their 2011 report, *Tracking the Sun IV*, that installed price for utility-scale photovoltaics ranged from $3.7 - $5.6 per Watt in the 2008-2010 period [Barbose 2011]. For the purpose of this calculation, a midpoint price of $4.65 is assumed for the year 2009.

2. **Determination of Energy Intensity:** The energy intensity of the US economy is a well known number. For the year 2009, energy intensity of the US economy was 7.52 MJ/$ [see chapter 2]. The US economy is primarily a service sector economy and the manufacturing of solar panels is an energy intensive activity. Hence, it is necessary to use a multiplier to estimate the energy intensity for the solar photovoltaics business. A multiplier of 3 is used for the purpose of this calculation. With this, the adjusted energy intensity is calculated to be 22.56 MJ/$.

3. **Calculation of Energy Input:** Energy input is calculated per Watt of installed capacity. Energy input in MJ is obtained by multiplying the price per Watt by the adjusted energy intensity. This is the primary energy input necessary per Watt of installed photovoltaic capacity.

4. **Calculation of Energy Delivered:** This calculation is based on capacity factor which was calculated earlier in this section. A capacity factor of 17%, which corresponds to Phoenix, AZ, is used for the calculation. The energy delivered is calculated in MJ per year per Watt of installed capacity.

5. **Calculation for Primary Energy Saved per Year:** Energy delivered by a solar panel saves fossil fuels. To calculate the savings in primary energy, an efficiency of 38% is assumed for thermal power plants that use coal. The energy delivered by the solar energy system is divided by this number to obtain the saving in primary energy. In other words, it is assumed that photovoltaic electricity replaces the use of coal for electricity production. This methodology is consistent with that used by BP Inc. for the calculation of primary energy in their annual report on world energy [BP 2011].

6. **EUF and EROEI Calculation:** EUF is calculated by dividing the calculated primary energy input by the yearly savings in primary energy. For the purpose of EROEI calculation, an operational life of 30 years with a negligible maintenance is assumed. EROEI is calculated by dividing the operational life in years by EUF.

The EUF and EROEI calculation is presented in Table 5.13. It is necessary to mention that the statistical sample size is quite small for deriving the installed cost of solar photovoltaic projects. This limitation has been mentioned in the paper referenced for the US data [Barbose 2011]. Prices of solar panels have declined since the year 2009. It is possible to purchase a panel for about $2 per Watt. However, the installed cost contains a number of items, apart from the panel cost. The total cost, and as a result,

EUF and EROEI, might have improved marginally in the year 2012. It is also interesting to analyze the impact of changing the multiplier used for adjusting energy intensity. A multiplier of 4 results in an EROEI of 3, while a multiplier of 2 yields an EROEI of 6.1 for USA. It is quite clear that although solar photovoltaics does return more energy than what it consumes, it is a source of energy with a low EROEI and high EUF.

Item	USA	India
Reference Year	2009	2009
Installed Cost	4.65 $/Watt	132 ₹/Watt
Energy Intensity (EI)	7.52 MJ/$	0.34 MJ/₹
Multiplier for EI	3	2
Energy Input per Watt	104.84 MJ	89.17 MJ
Energy Output per Watt	4.08 Wh/Day	3.87 Wh/Day
Yearly Energy Output per Watt	5.36 MJ	5.09 MJ
TPE Saved per Year per Watt	14.11 MJ	13.40 MJ
EUF	7.4	6.7
EROEI	4	4.5

Table 5.13: EUF and EROEI Calculation for Photovoltaics

The table also shows the EROEI calculations for photovoltaics in India. Once again, the sample size is very small for determining the price. Secondly, a multiplier of 2 is used to adjust the energy intensity, since the percentage of service sector in GDP is much less for India than that for the US. The calculation for energy output assumes a capacity factor of about 16% for India. On the basis of these assumptions, the EROEI for solar photovoltaic electricity in India is calculated to be 4.5. It can be observed that the numbers for EROEI and EUF are not too different for India and USA.

Dispatchability and Predictability of Solar Photovoltaic Energy

Just like the wind energy, solar photovoltaic energy is not a dispatchable source of energy. Energy is delivered only as long as the sun is shining, irrespective of the demand from the electric grid. However, this source of energy is much more predictable as compared to the wind energy. The timings for sunrise and sunset are known in advance, and even on extremely rainy days, at least some energy will be delivered, thanks to the diffused solar radiation. In India, monsoon season lasts from June to September. During the remaining 8 months, ample sunshine is available with the probability of overcast days being quite low in most parts of the country. Conversely, in many parts of India, during the months of July and August, due to continually overcast conditions, the availability of direct solar radiation is less than the diffused radiation. The solar insolation reaching earth is itself almost a constant. This predictability certainly helps in operating an electric grid.

Summary of Benchmarks for Photovoltaic Power

Item	Value
Capacity Factor	15% to 18%
Dispatchability	Not dispatchable
Peak Power Density	80 - 200 W/m^2
Energy Density	100 - 300 kWh/m^2/year
EUF	5 to 7.5
EROEI	4 to 6
Cost	High

Table 5.14: Energy Benchmarks for Solar Photovoltaic Energy

Note: The calculations in Table 5.14 do not take into account locations such as Seattle, USA, which are unsuitable for a solar photovoltaic installation.

Solar Tracking Systems

Photovoltaic power suffers from a poor capacity factor. One of the reasons for the low capacity factor is the movement of sun in the sky, resulting in the COS(θ) effect. The reduction in energy output due to the COS(θ) effect can be avoided by following the sun as it traverses the sky. A solar tracking system achieves this purpose. Such systems are also called heliostats. With such a system, it is guaranteed that the sunlight will always be incident perpendicular to the solar panel, generating maximum energy output. Solar tracking can either be performed using a single axis tracker or a dual axis tracker. A dual axis tracking system will perfectly track the sun in all seasons. A single axis tracking system tracks the daily movement of the sun, but it cannot compensate for the seasonal movement. Single axis tracking is usually sufficient for photovoltaic panels.

Optical Concentration and Multi-junction Solar Cells

Tracking systems are sometimes used along with optical concentrators to improve the energy output of solar cells. An optical concentrator is generally built using a Fresnel lens or mirrors. Both these devices are relatively inexpensive as compared to a solar cell. The idea is to improve the utilization of expensive solar cells by increasing the solar insolation on individual cells, with the use of inexpensive optics. An optical concentrator collects sunlight over an area that is larger than the solar cell which results in an increase in the effective insolation received by the solar cell. Sometimes, such systems make use of the more expensive, but highly efficient, multi-junction solar cells.

An important characteristic of such systems is the concentration ratio. It is the ratio of area over which sunlight is collected to the area of solar cell. In general, a lens based system provides a higher concentration ratio as compared to a mirror based system. However, for a high concentration ratio, the tracking of the sun needs to be perfect. There are a few of challenges in making such a system successful:

- Special arrangements are necessary for cooling the solar cells to prevent damage due to excessive heat.
- Solar cells deliver lower power at high temperatures.
- Optics and tracking needs to be perfect to ensure that each solar cell in the panel receives the same insolation. A mismatch leads to a reduced power output. Also, a mismatch may lead to hot-spots that result in damage to solar cells.
- Tracking implies moving parts, control hardware, and software. A conventional solar panel based system requires almost no maintenance, except cleaning the surface. A system that involves tracking is expected to require at least some maintenance over the years.

Though tracking and concentration seem to offer a lot of promise, there are only a few systems that are available commercially. It is doubtful if these systems provide a better return on money or energy invested, as compared to the ordinary photovoltaic systems.

Use of Batteries with Photovoltaic Systems

A battery is a device that stores energy. The stored energy can be used whenever there is demand. Solar photovoltaic power is not a dispatchable source of energy. The use of batteries, at least to some extent, solves the problem of dispatchability. For instance, most of the household electricity consumption takes place in the evening and night, while the photovoltaic electricity is generated only during the daytime. Batteries are used to store energy during the daytime and this stored energy is then used during the evening and night. Most of the rural and off-grid uses of photovoltaic electricity involve the use of batteries. Use of lead-acid batteries is common. The commonly available car battery is a lead-acid battery. Solar energy applications normally use a variant of lead-acid batteries, known as the deep-discharge battery. In a solar energy application, the battery is expected to charge during the course of the day and the battery delivers electricity during the few hours of evening and night. The charging and discharging takes place at lower currents as compared to the car battery. The battery is expected to loose most of the charge every evening, requiring a recharge the next day. These characteristics require a different design of battery as compared to the car battery. There are at least a few problems in using batteries in a photovoltaic application:

- Energy efficiency for lead-acid batteries is only about 70%. This reduces the available energy further. A photovoltaic panel is an expensive device. Loosing 30% of the energy generated by the panel is a rather large loss.
- A lead-acid battery can have a useful life of about 2 to 4 years, assuming daily use. This is rather short, compared to a maintenance-free life of 25 to 30 years for solar panels. Some types of batteries also require regular maintenance, which is quite problematic in many applications.
- Cost of lead-acid batteries is quite high. In July 2011, battery prices were about $213 per kWh of storage capacity in the US, and about ₹6,000 per kWh in India. While batteries increase the cost of solar photovoltaic systems, energy output is reduced by 30% due to the impact of battery efficiency. As a result, a solar photovoltaic system with batteries takes even longer to return the energy that was used in manufacture, installation, and maintenance of the system. According to my calculations, such a system installed at an average location in India would take more than 8 years to return any net energy. However, the battery is expected to last not more than 2 to 4 years. Hence, it is doubtful if such systems will return any net energy.

- Lead-acid batteries are quite heavy, requiring well designed storage spaces for large arrays of batteries that are necessary for large systems.
- Nickel Metal Hydride (NiMH) and Lithium-Ion are more modern batteries, that are quite expensive. Though these types of batteries have longer life, higher energy density, and better efficiency, their use in solar energy applications is limited, primarily due to cost.

Rural Photovoltaic Applications: Problems of Maintenance

In India, there are frequent news-reports about lighting up of a remote village, or street, with solar photovoltaic lighting. These systems consists of a small solar panel, along with a lead-acid battery and CFL, or more recently, LED lights. Each day, the solar panel charges the lead-acid battery during the daytime, and the stored energy is used in the evening and night to light up homes and roads. Most of these locations are remote, with no access to the electric grid. Not surprisingly, the villages, where such systems are employed, are among the poorest. The Jawaharlal Nehru Solar Mission in India proposes a 1000 MW installed capacity for off-grid rural systems by the year 2017 [Solar India]. A majority of these systems will be installed with government funding, since the residents of the remote villages and village governments are unlikely to have the financial power to purchase the systems at market price. These systems tend to work very well for a couple of years. The challenge will come once the batteries near the end of life. The 1000 MW capacity implies about 40 million systems, at 25 W per system. A battery capacity of 100 Wh per system would be a reasonable assumption. The battery capacity necessary for 40 million systems would be about 4 million kWh. Even at ₹5,000 per kWh, the battery cost would be 20 billion rupees. This expense must repeat every three years, which is the expected battery life. Recycling of used batteries will only result in a small savings in this expense, and funding such an expense will be a significant challenge. Another challenge would be logistical in nature. These systems would be spread out across the country, in remote places. Reaching out to these communities to carry out regular maintenance, or replacement of batteries, is quite difficult. These observations lead to the following conclusions:

- The main advantage of a solar panel is long life with almost no maintenance requirement. This advantage is negated by a system that employs lead-acid batteries.

Illustration 5.6: Solar Powered Lamp on the trek to the Sinhagad Fort near Pune, India.

- A significant amount of research is required in battery technology to develop batteries that would have a life matching that of the solar panel. The price for such batteries needs to be low in order to be cost effective. Energy storage based on super-capacitor technology offers promise as far as long, maintenance-free life is concerned. However, the cost of this technology is prohibitively high, at present.
- The off-grid part of the Jawaharlal Nehru Solar Mission is unlikely to be successful without technological advances in the battery technology.

Hurdles in Transition to Solar Photovoltaic Electricity

The solar insolation that reaches the earth is more than sufficient to fulfill all the energy requirements of our civilization. This was illustrated earlier in this chapter with India as an example. Even if only 1% of the land area is used to harness solar energy with photovoltaics, it will be possible to satisfy the entire energy demand of the country. What are the hurdles in making this transition possible? Apart from the obvious hurdle of high capital cost, some of the problems are generic and technological, while others are economic:

1. Photovoltaic electricity is not dispatchable. It is not possible to adapt the energy supply based on the demand for electricity from the grid. As a result, a large scale use of photovoltaic electricity is not possible without the use of batteries, or some other energy storage technology, which allows for some dispatchability.

2. The existing battery technology is too expensive and batteries have a short life. There is a dire requirement for research in battery technology to develop batteries that are inexpensive and have a life comparable to that of the solar panels. One possibility is to make use of pumped storage systems discussed earlier in this chapter. While these system have a very long life, their capital costs add to the already high capital cost of photovoltaic electricity. The amount of pumped storage available is also limited by the suitability of geography.

3. A capacity factor of 16 to 17% and the absence of suitable storage systems implies that some alternative arrangement must exist when photovoltaic electricity is not available. Connectivity to a large electric grid certainly helps. However, some alternative to solar power needs to be in-place, on the grid. One possibility is to make use of hydroelectricity, when photovoltaic electricity is not available. However, the hydroelectric potential in most countries is no match to the potential for photovoltaic electricity generation. This limits the use of hydroelectricity as a backup on the grid.

4. The poor EUF and EROEI numbers for solar photovoltaic electricity indicate that such systems require a large energy input. In a world of dwindling energy resources, this could become a challenge, as a very large scale production of such systems may have an impact on the energy available to produce these systems, as well as on the cost of energy. A

corollary to this statement is that the production of photovoltaic panels cannot be increased very rapidly without increasing the total energy consumption. This situation can change only with a technological breakthrough.

5. High capital costs and low energy density imply that only infertile land with low water availability may offer better return on investment for the generation of photovoltaic electricity, as compared to agricultural or industrial use. This factor will be of importance when planning for very large scale solar energy farms.

Solar Thermal Electricity

Man has always known that high temperatures can be obtained by concentrating sun rays. Legend tells us that in ancient Greece, Archimedes used mirrors to concentrate sunlight on enemy ships which caused them to burn and sink. Solar thermal electricity generation essentially uses the same principles. A large number of mirrors are used to concentrate sunlight on a small target. The high temperature so obtained is used to boil water to generate steam at very high temperature and pressure. The steam is used to drive conventional turbines to generate electricity. There are a couple of types of solar thermal electricity systems, but the solar thermal power tower is the most promising one.

Solar Thermal Power Tower

This system actually consists of a tower, on which sunlight is focused using a large number of mirrors. The mirrors are spread over a large area, and track the movement of sun in the sky. The mirrors are very similar to the ordinary flat mirrors. They are controlled by motors, electronics, and software to track the sun to direct sun rays on the tower at all times. These sun-tracking mirrors are called heliostats. In a solar thermal electric system, thousands of heliostats concentrate sunlight on the central tower, which efficiently collects the heat to generate steam at high temperature and pressure. This makes it possible to efficiently drive a steam turbine to generate electricity. One advantage of a solar thermal tower system is that most components of the system are based on established technology. It uses flat mirrors, which are inexpensive to manufacture as compared to parabolic reflectors. The tower directly heats water to generate steam and does not require any other liquids to carry the heat. Technology for steam turbines is well developed owing to their use in modern thermal

Illustration 5.7: Solar Thermal Electric Power: Heliostats and Tower. Credits:
paulrommer/Shutterstock.com

power stations. The complexity is in the accurate tracking of sun by heliostats placed on an uneven terrain, and in the efficient collection of concentrated heat at the tower. However, with advances in electronics, and software technology, the heliostat should not be a particularly expensive, or complex component, anymore.

There are only a few such systems in the world. The most notable one is still in the works at Ivanpah, in the Mojave desert, in California. The Ivanpah system uses technology developed by BrightSource Energy of Oakland, CA. This is a 377 MWe system, spread over an area of about 1,400 hectares. It is perhaps the largest solar-electric system currently under construction, employing over 300,000 heliostats [BrightSource 2012].

Estimation of Energy Generation by the Ivanpah Plant

The calculations in this section are based on publicly available data. BrightSource Energy, or any other participants of the Ivanpah project, have not reviewed, verified or endorsed these calculations.

Ivanpah is a great location for a solar-electric systems. Data from NASA indicates that the annual average Direct Normal Insolation (DNI) at Ivanpah is about 6.95 kWh/m²/day. Compare this with about 5.31 kWh/m²/day for Pune, in India! The Ivanpah system is rated at 377 MWe. This must be the net peak power output when the solar insolation is at 1000 W/m². If this were a constant power output from the plant for 12 hours every day, the energy generated per year would have been 1,651,260 MWh. However, as sun traverses the sky, the amount of insolation reaching the earth changes due to the difference in the length of atmosphere through which the sunlight must pass. When the sun is near the horizon, insolation is at the lowest. Moreover, the number of hours of sunshine varies through the seasons. Cloud cover too changes from season to season. Instead of actually going into the calculations of air-mass and other factors to estimate the insolation, it is easier to simply use the NASA data for DNI. The annual average for DNI not only takes into account the daily traversal of sun in the sky but also the seasonal changes and the weather. At a constant insolation of 1000 W/m², average daily insolation would have been 12 kWh/m²/day. However, Ivanpah is reported to have an annual average insolation of 6.95 kWh/m²/day [NASA EOS]. Thus, a down-rating factor of 6.95/12 should be applied to estimate the energy output. This number will take care of the traversal of sun, seasonal changes, and weather. Another 10% down-rating is necessary to take into account the fact that as the sun traverses the sky, the amount of sunlight incident on the mirrors changes due the change in the angle of incidence of sunlight on mirrors. Since the mirrors need to reflect the sunlight onto the solar tower, the mirrors cannot remain perpendicular to the sunlight at all times. After applying the down-rating factors, the estimated energy output from the Ivanpah power plant is calculated to be 861,164 MWh per year, or about 3.1 million GJ. Note that this calculation does not include any down-rating factors specific to the BrightSource technology. For instance, the turbine efficiency will depend on the temperature and pressure of the steam. In the early morning and evening hours, steam temperature will be lower due to the lower solar insolation. This can result in a reduced efficiency. It is possible to overcome this problem by using a backup such as natural gas to obtain a stable steam temperature and pressure. Hence, it needs to be stressed that the calculations in this section could be somewhat aggressive and the actual output could be lower. The electricity generation data from the Ivanpah plant will be watched with a lot of interest, once this plant becomes operational.

Calculation of Benchmarks for Solar Thermal Electricity

Capacity Factor:

This is the actual yearly energy output divided by the energy output for generation at peak power for the whole year. The previous calculation for the Ivanpah power plant estimates that the actual energy output is expected to be about 3.1 million GJ (electric) per year. The peak energy output (E_p) is obtained if the peak power output of 377 MW is available for the entire year:

$$Ep = 377 \times 24 \times 365 \times 3.6 = 11,889,072 \ GJ$$

$$Capacity \ Factor = \frac{3,100,000}{11,889,072} \times 100 = 26 \ \%$$

This is a fairly reasonable capacity factor for an intermittent source of energy.

Peak Power Density:

Peak Power Density is the peak power output divided by the total area that the system occupies in square meters. The Ivanpah project is planned to be spread over an area of 1400 hectares. The peak power density is calculated to be:

$$peak \ power \ density = \frac{(377 \times 1000000)}{(1400 \times 10000)} = 27 \ W/sq.mtr$$

It can be observed that the peak power density is much less than the photovoltaic electricity.

Energy Density:

Energy Density is the amount of energy generated in a year per square meter of land. The energy output of 861,164 MWh per year is generated over an area of 1400 hectares. The energy density is calculated to be:

Even the energy density is much less than the photovoltaic electricity.

$$energy \ density = \frac{861,164,000}{(1,400 * 10,000)} = 62 \ kWh/sq.mtr/year$$

Total Primary Energy (TPE) Saved per year:

The Ivanpah project is expected to generate about 3.1 million GJ of energy per year. The TPE savings are calculated by assuming that the same amount of energy is generated using a coal based power plant with an efficiency of 38%. The energy content of coal required to produce 3.1 million GJ of electrical output is the savings in TPE:

$$TPE \ Saved = \frac{3.1}{0.38} = 8.16 \ million \ GJ$$

EUF and EROEI for Solar Thermal Electric Generation:

The calculation of EUF and EROEI depends on the cost of the system and on the energy generation by the system. While energy generation can be estimated fairly easily, it is rather difficult to estimate the *reasonable* cost for solar thermal electric systems. The number of such systems implemented in the world are very few. These systems do not benefit from economy of scale that photovoltaics does and the sample size for cost data is just too small. Yet, calculation of EUF and EROEI is attempted for the Ivanpah system based on some publicly available data to illustrate the immense potential that these systems could have in the future.

Energy spent upfront on the Ivanpah project can be estimated on the basis of the energy intensity of the US economy and the total cost of the Ivanpah project. The energy intensity of the US economy was about 7.5 MJ/$ in the year 2009. In the year 2011, after adjustment for inflation, the energy intensity is unlikely to change much. However, it is expected that some of the activities involved in the manufacturing and installation of a solar thermal tower will be more energy intensive. It is assumed that the Ivanpah project has double the energy intensity of the US economy, implying an energy intensity of 15 MJ/$. The cost of the Ivanpah project is estimated to be about 2.18 billion dollars. Thus, the energy input for the Ivanpah project can be calculated to be about 32 million GJ. This is the Total Primary Energy (TPE) input. It was calculated earlier that the TPE saved per year due to energy generated by the Ivanpah plant is estimated to be 8.16 million GJ. Thus, the EUF for the Ivanpah solar power tower is expected to be about 3.9. In other words, in a little under four years, the Ivanpah solar power tower will return the energy spent in its making. Assuming a life span of 30 years, the Ivanpah solar power tower will return about 7.7 times the energy input, assuming that the energy required for operations and maintenance is negligible compared to the annual energy generation. There is also an assumption that the land is available at a very low price. This assumption is valid in desert or arid areas, where the land cannot produce anything else. Such locations typically have a high solar insolation and are very suitable to produce electricity. It needs to be mentioned once again that these calculations are based on several assumptions. If and when a large number of such systems are built, a better estimate of EUF and EROEI should be possible.

Summary of Benchmarks for Solar Thermal Power Tower:

These benchmarks are the estimates for the Ivanpah system. They benefit from the fact that Ivanpah is an excellent location for solar thermal power. These numbers are not endorsed or verified by BrightSource Energy, or any other participants in the implementation of the Ivanpah system.

Item	Value
Capacity Factor	26%
Dispatchability	Not Dispatchable
Peak Power Density	27 W/m^2
Energy Density	62 kWh/m^2/year
EUF	3.9
EROEI	7.7

Table 5.15: Summary of Benchmarks for Solar Thermal Power Tower

Solar Thermal Power Generation Estimates for India

It is interesting to know that most places in India receive a lower average Direct Normal Insolation (DNI) as compared to Ivanpah, California. Compare the yearly average DNI of 5.31 kWh/m^2/day for Pune, and 5.69 kWh/m^2/day for Jaisalmer, with 6.95 kWh/m^2/day for Ivanpah. The main reason for the lower DNI is the extreme cloud cover during the monsoon season. For Pune, the average DNI during the month of July is 1.77 kWh/m^2/day, while in January, it is 7.14 kWh/m^2/day, though the number of daylight hours are considerably less in January. In fact, for Pune, in July and August, the diffuse radiation is higher than DNI [NASA EOS].

It is reasonable to expect that the yearly energy output of a solar thermal power tower system is directly proportional to the DNI. Hence, the factors that depend on the annual energy output can be be estimated by scaling down the numbers for the Ivanpah system by the DNI ratio. Hence, the capacity factor and energy density of such a system in Pune will need to be scaled down by 5.31/6.95. Energy density and capacity factor are calculated to be 47 kWh/m^2/year and 20%, respectively. To calculate EUF and EROEI, it is necessary to know the cost of a solar thermal power system in India. Unfortunately, no such system is presently implemented, and only a guess can be made. The bulk of the cost of the system is in components that can be manufactured locally in India. Even the control software for heliostats and electronics can be developed locally. Labor costs are much lower in India than in California. Thus, it should be possible to manufacture and install a solar thermal power tower at a much lower cost in India. Generally, the lower cost will also imply a lower energy input. In effect, in spite of a slightly lower energy output, EUF and EROEI for a solar thermal power tower in India should be very similar to that in California.

Land Requirement for Solar Thermal Power to Satisfy the Electricity Requirement in India

It is necessary to make certain assumptions for this calculation. Firstly, the target year is assumed to be 2030 to build a sufficient number of systems to meet the total electricity demand. It is assumed that the population of India will stand at 1.3 billion in the year 2030, and the per capita electricity usage will be at 1000 kWh per year. This implies a requirement of 1300 billion kWh of electricity per year. A solar thermal power tower in India is expected to generate electricity at an energy density of about 47 kWh/m^2/year, which is the same as 47 million kWh per square kilometer per year. Thus, to generate 1300 billion kWh every year, land requirement is calculated to be 27,660 km^2. While this seems like a large number, it is only about 0.8% of the total area that India occupies. Compare this with the land used for agriculture, which is about 1,790,000 km^2. Even the fallow land in India occupies an area of 240,000 km^2 [UN Data]. Food and clothing are our basic needs and to satisfy them over 50% of the total land area is utilized. Likewise, a reasonable supply of electricity is the need of our civilization. Utilizing only about 1% of the total area for electricity generation should not be an unacceptable proposition.

Advantages and Disadvantages of Solar Thermal Electricity

Advantages:

- Solar radiation is optically concentrated on the collecting surface to boil water. This method yields a much better efficiency as compared to using a liquid to collect and transfer heat to a boiler.

- A solar thermal power tower system can make use of available terrain. The terrain need not be perfectly flat. It should be possible to design heliostats that are lightweight, so that a heavy foundation is not required for installation. Less work on the terrain not only implies lower costs but also a lower energy input.

- Desert and arid areas are perfectly suited for such a system. Installing such a system allows for the productive use of land, which is otherwise unsuitable for agriculture.

- Small trees, shrubs, and small animals can coexist with heliostats. Very little environmental impact is expected due to such a system.

- The individual components used to build this system do not use very unique, or new technology. Mirrors, boilers, and steam turbines have always been available. Even heliostats are not very difficult to design and manufacture, considering the advances in computer hardware and software technology.

- Perfect tracking of the sun results in better collection of heat energy.

Disadvantages:

- Low capacity factor and no dispatchability.
- Low energy density, requiring relatively large land use.
- High capital cost, though it should be lower than photovoltaic electricity when systems are mass produced.
- These systems may require a lot of water, even when recycling is implemented. Water use can be a problem since such systems are likely to be located in desert and arid regions.

Heat Storage to Improve Capacity Factor

The amount of solar insolation at a location puts an upper limit on the deliverable energy. However, capacity factor can be improved by employing energy storage techniques. It is possible to store heat using molten salts. In such a system, while a part of the heat is used to generate steam, a portion is used to heat salts. Typically, a mixture of sodium nitrate and potassium nitrate is used. During cloudy periods of the day, or after sunset, heat stored in the salt can be used to convert water into steam to drive turbines. In other words, electricity can be generated on demand, leading to an improvement in capacity factor. An article in the Scientific American magazine claimed that round trip efficiency of a molten salt storage system could be as high as 93% [Biello 2009]. In other words, if electricity is produced using the heat stored in molten salt, there will only be a 7% reduction in the net output, as compared to direct production of electricity. The cost of such an energy storage system is expected to be only about $50 per kWh, according to the same article. Compare these numbers with the efficiency of 70% for a lead-acid battery, and a cost of about $200 per kWh. This data implies that there will be a minimal increase in the price of electricity as a result of the use of the molten salt storage, while the capacity factor can be improved significantly. It is necessary to mention though, that the molten salt storage technology, as applied to solar-thermal electricity, is not field-proven on a sufficiently large scale. Nevertheless, it is a good candidate for research and development.

Comparing Solar-Thermal and Photovoltaic Electricity

- Dual-axis tracking is mandatory for solar thermal electric power, while most photovoltaic systems do not use tracking. As a result of tracking, capacity factor is higher for solar thermal electricity, while energy density is lower.
- Lower energy density for solar thermal electricity implies higher land-use for the same output energy.
- Improving capacity factor for photovoltaic electricity with the use of batteries is a very expensive proposition, and such systems may not return any net energy. Molten salt heat storage is a very promising technique to improve capacity factor for solar thermal electricity.
- Photovoltaic electricity is a more democratic energy source, as small systems can be implemented on individual rooftops. Solar thermal electricity requires large, centralized systems.
- Photovoltaic electricity is a very elegant method of energy conversion with no moving parts and almost no maintenance requirements.
- While it may seem that solar thermal electricity is at least as expensive as photovoltaic electricity, there is reason to believe that mass production can bring down the costs significantly for solar thermal electricity.

Low Temperature Applications of Solar Thermal Energy

Solar Water Heating:

Solar water heating is the most widely deployed application of solar energy in India. Yet, less than 1% households may have solar water heaters installed. Most middle-class households either use electricity or gas-based water heating. Lower income households use firewood or other forms of biomass. Most parts of India have ample sunshine for 8 to 10 months of the year, and using solar energy is the most environment-friendly option for water heating. This particular application of solar energy is commercially viable. However, high capital cost of solar water heaters is a hurdle.

Illustration 5.8: Solar Water Heater. Credits: Pavel Ganchev/Shutterstock.com

Solar water heaters typically use the thermo-siphon technique (see Illustration 5.8). An insulated storage tank is connected to a collector panel that absorbs sunlight to heat water. Heated water rises and enters the storage tank. Cooler water from the storage tank is displaced and enters the collector panel to get heated. In this process, temperature of the water in the storage tank continues to rise during the day, as solar radiation is absorbed by the collector. The storage tank is well insulated with injected

Polyurethane Foam (PUF) to prevent the loss of heat during the night. The collector panel generally consists of evacuated tubes with heat absorber coating on the inner tube. Such systems have a thermal efficiency of 70% or more. Solar water heaters are sometimes fitted with an electric heating element. This serves as a backup during the monsoon season, when solar insolation reaching the ground drops significantly. These systems are perfect for regions such as India, where temperatures never go below the freezing point at most locations. In colder climatic regions, it becomes necessary to prevent the water in the tubes from freezing. This takes a toll on the overall efficiency.

It is fairly easy to estimate the EUF and EROEI for solar water heating in India. Methodology of calculation will be similar to that used for other forms of renewable energy. According to individual experience, about 10 to 60% of domestic electricity consumption takes place for the purpose of heating water. There is a significant seasonal variation in consumption. For instance, a typical family of 4 in the city of Pune, India, would consume about 4 kWh/day for heating water in the month of December. The same family would consume only about 1 kWh/day in April, due to the hot summer weather. It is estimated that a solar water heater with a capacity of 100 liters will save 860 kWh of electricity per year. This data assumes a family size of 4 and a weather pattern of Pune, India. The cost of such a water heater was about ₹20,000/- in the year 2011. A saving of 860 kWh corresponds to a savings in the primary energy input of about 3,400 kWh, considering the typical efficiency of thermal power plants and the savings in electrical transmission and distribution losses. The price of ₹20,000/- corresponds to a primary energy input of about 3,700 kWh, assuming that a solar water heater has twice the energy intensity as compared to the Indian economy. Thus, a solar water heater is expected to return the energy used in its manufacture and installation in just 13 months. Solar water heaters can be expected to operate for at least 20 years without any maintenance, implying an EROEI of 18!

The savings of 860 kWh of electricity per year corresponds to a saving of about ₹ 4,300 per year, assuming an electricity tariff of ₹ 5/kWh. On an initial investment of ₹ 20,000/-, the yearly return works out to be 21.5%. Depreciation in the value of the water heater will be more than compensated by the increasing price of electricity.

Quite clearly, the solar water heating application makes sense as far as economics and energy returns are concerned in a tropical country like India. The only hurdles are the capital cost and the attitude of builders and housing complexes towards their installation and use. Hopefully, these problems can be solved by the Jawaharlal Nehru Solar Mission by making capital available, and by enacting laws that remove the hurdles [Solar India].

Solar Cookers:

A solar cooker typically consists of a box covered with glass. The box is lined with reflecting material on the inside. It may also have reflecting flaps which direct sunlight into the box. This construction concentrates sunlight on the food container inside the box. Most solar cookers are inexpensive devices that do not implement solar tracking. As a result, the maximum temperature that can be reached inside the cooker is less as compared to the traditional cooking methods. A low temperature cooking application is bound to take longer cooking time. Though solar cookers have been around for decades, this application never really took off, in spite of being quite inexpensive. Solar cooking is a slow process that does not match the lifestyle of most people, even in villages.

Biomass and Biofuels as Alternative Sources of Energy

Hundreds of millions of poor people in the developing nations use biomass as the only source of energy for their day-to-day needs. As such, biomass cannot be termed as an an *alternative* source! In fact, biomass was the primary source of energy before the beginning of the fossil fuel era. In general, biomass in the solid form is not a clean burning fuel. The particulate matter released on combustion of biomass is a major source of pollution in developing nations. The age of fossil fuels allowed millions to migrate from biomass to clean-burning, fossil fuel derivatives. Yet, biomass is considered to be an alternative source of energy as sources of fossil fuels dwindle or get more expensive.

Biomass generally takes one of the following forms:

- Firewood gathered from forests, fields, cities, etc.
- Crop residues which are not used as animal feed or burnt in the field.
- Animal waste such as cow dung, either directly used as fuel, or converted into biogas.
- Charcoal obtained from wood.
- Biodiesel obtained from oil seeds.
- Alcohol (ethanol/methanol) obtained from sugar, corn, cellulose, etc.
- Municipal waste, either used directly, or after conversion into biogas.

Plant-derived alcohol and biodiesel are also known as biofuels. The only difference between biomass and biofuels is the commercial plantation that biofuels are associated with.

Biomass Applications and Usage

Solid biomass is used in stoves, for cooking and other domestic heating applications. Such household use is common in developing nations, especially among poor sections of the society. Biomass is sometimes used to directly produce electricity. For instance, sugar factories use bagasse to generate electricity. According to the UN database, in the year 2008, biomass was used to obtain about 1.2 million TJ of energy in India. This was about 6% of the total primary energy (TPE) consumption. In the same year, the crude oil consumption in India was equivalent to about 6.4 million TJ. In most of the developed world, the use of solid biomass in domestic heating applications is negligibly low. However, there has been a renewed interest in biofuels in the recent times due to the high price of crude oil and environmental concerns. It is common in many countries to add a small percentage of ethanol to gasoline. Similarly, biodiesel obtained from oil seeds and other sources can be used to replace the use of conventional diesel in vehicles. There is hope, that at some future date, biofuels can replace the fuels obtained from crude oil. United States is the largest producer of biofuels. In the year 2011, the US production of biofuels stood at 28 MTOE, which was about 48% of the world total of 58.9 MTOE. During the same year, crude oil production in the US was 352 million tonnes, while the world production was at 3996 million tonnes [BP 2012]. Most of the biofuel production consisted of ethanol. Quite clearly, biofuels have a long way to go before making any impact on the crude oil market.

An important impact of high price of crude oil is the possibility that millions of poor people in developing nations will move away from LPG and kerosene to use biomass in some form. In India, kerosene and LPG are subsidized by the government. High price of crude oil and oil-derivatives have made it very difficult to sustain subsidies in a difficult economic situation. Most poor people will find it difficult to afford the market price of kerosene and LPG, once the subsidies are removed. The only option for many will be to switch to biomass in some form. This is another reason that the consumption of biomass can significantly increase. As a result, it is necessary to examine the use of biomass as a renewable source of energy, and to analyze the capacity of biomass to fulfill the energy demands of mankind.

Biomass as a Renewable Resource

It is normally claimed that biomass is a renewable resource. Is this a fact? A resource can be classified as renewable if it is replenished after use, and the net supply does not decrease over a reasonable timespan. Firewood gathered from forests is a renewable resource, provided that the activity of gathering does not involve cutting of productive trees. In a balanced ecosystem, nutrients absorbed from soil are returned back to it through natural processes. A forest may thrive without external inputs such as fertilizers, or man-made irrigation systems. Using charcoal or firewood gathered from such a forest is certainly a renewable use of resources.

Crop residue and biogas are biomass resources which utilize the heat content of materials which otherwise would have been discarded. Crop residue, in general, is not a renewable resource since farming methods are not organic/sustainable, in most cases. The real question about these resources is their ability to scale to meet a useful portion of energy demand. The energy density of crop residues is certain to be very low. This resource is widely dispersed, and creating a large scale supply of fuel (a concentrated source of energy) with crop residues is not possible without a significant energy input. As a result, the net energy return is likely to be only marginally positive. Biogas plants that make use of municipal waste have existed for a long time. However, they have failed to contribute more than a small portion of energy supply in towns and cities. In summary, while using heat content of crop residues and municipal wastes is a good idea, their net contribution to the total energy supply will always remain low.

Energy crops such as sugarcane and corn are heavily dependent on the use of chemical fertilizers to replenish the nutrients taken from soil. Chemical fertilizers are processed from natural gas. These crops also require a large quantity of water which requires pumps that use electricity, which is primarily obtained from fossil fuels. Other farming activities too are dependent on machinery that runs on fossil fuels. Such crops, cannot be classified as renewable due to the dependence on fossil fuels. In fact, it is questionable how much net energy corn or sugarcane based ethanol returns.

Another important energy resource is the plantations of inedible oil seeds which are used to produce biodiesel. In India, there has been significant interest in the Jatropha tree plantation in the recent years. This tree produces oil seeds which are processed to generate biodiesel. Jatropha can be planted even in marginal land and does not require an excessive amount of water. Such a plantation, which uses hardly any external inputs, can be termed as a renewable resource. It is claimed that using Jatropha oil seeds is a very productive method of biodiesel generation.

Calculations for Energy Benchmarks for Biofuels

The calculations in this section are specific to biodiesel produced from Jatropha seeds. Energy benchmarks for other biofuels are specified based on quoted references. There is a lot of interest in Jatropha plantation due to the promise of high yield per hectare and high economic returns. Jatropha plantation can yield about 3,000 kg of biodiesel per hectare, per year [CJP 2012]. This assumes good land quality and a sufficient availability of water.

Energy Density:

One kg of biodiesel contains approximately 42 MJ of energy. Thus, a production of 3,000 kg/hectare/year is equivalent to an energy availability of 126,000 MJ/hectare/year. Using $kWh/m^2/year$ as units, energy density is calculated to be 3.5 $kWh/m^2/year$. This number is the *gross* energy density. The *gross* energy density does not include any energy inputs for the Jatropha plantation or the energy spent as a part of manufacture and distribution of biodiesel. Compare this with the net energy density of 100 to 300 $kWh/m^2/year$ for photovoltaic power. The difference is almost two orders of magnitudes!

It will also be useful to calculate the energy density for corn-based ethanol. It is assumed that the energy content of corn seed as well as all other parts of the maize harvest are utilized for energy production. The energy density of dried plant is assumed to be 15 MJ/kg and the production of maize is assumed to be as high as 20 tonnes/hectare/year. The energy available is calculated to be 300,000 MJ/hectare/year, or 30 $MJ/m^2/year$, equivalent to about 8 $kWh/m^2/year$. Note that this is the gross energy density. Assuming an EROEI of 2, which is much more than what many studies report, the net energy density is calculated to be only 4 $kWh/m^2/year$.

Power Density:

Gross power density is calculated by dividing the energy density in $MJ/m^2/year$ by the number of seconds in a year. This number is calculated to be only 0.4 W/m^2. The net power density will be even lower. Compare this with the power density of 80 to 200 W/m^2 for photovoltaic power.

EUF:

It is well known that Jatropha grows in marginal land with very little effort for land preparation. Such a plantation would require a minimum infrastructure expense. Since capital expense for such a plantation is very low, it follows that EUF is insignificantly low.

EROEI:

There are only a few inputs that the production of Jatropha biodiesel requires. In India, Jatropha plantation is viewed as a means of poverty alleviation through employment creation in rural areas. As a result, the use of farming machinery is deemed unnecessary and the use of manual labor is preferred for Jatropha farming. It is estimated that 1 hectare of Jatropha plantation generates about 300 man-days worth of work [CJP 2012]. Thus, the main energy input for Jatropha biodiesel production corresponds to the resource consumption of a subsistence farmer in rural India. Quite obviously, hardly any fossil fuel resources can be counted as input for the production of Jatropha biodiesel. This discussion assumes that biodiesel is used close to the point of production, and transportation and marketing expenses are minimal. With these assumptions, the EROEI for Jatropha biodiesel can be claimed to be very high. This is very similar to the EROEI for firewood gathered manually from forests. For such biomass resources, EROEI calculation is unnecessary as the inputs are derived from renewable sources. This conclusion also implies that the net energy density for Jatropha biodiesel will be equal to the gross power density.

EROEI for ethanol derived from corn and sugarcane is expected to be very low due to the nature of inputs that such plantations require. Prof. Vaclav Smil mentions that the net energy return for ethanol produced from corn and sugarcane in the US is only marginally positive, while ethanol produced from sugarcane in Brazil has a better energy return, implying a higher EROEI [Smil 2005]. In either case, the EROEI is less than 2.

Biomass: Advantages

There are some advantages of biomass over other renewable resources:

- Biomass requires low capital investment and hence has a low value for EUF. This is particularly attractive for any developing nation, where capital is hard to find.
- Biomass is highly accessible to even the poorest people. It is the most democratically distributed energy resource.

- Biomass and its derivatives can directly replace solid, liquid, and gaseous fossil fuels. Other renewable resources are mainly used to produce electricity.
- Jatropha biodiesel has a good EROEI and can be a useful fuel source in rural areas, owing to its liquid nature.
- Biofuels, particularly biodiesel obtained from plants like Jatropha, have an added benefit of creating employment in rural areas on a large scale.

Biomass: The Problem of Scalability

A major disadvantage of all types of biomass is the extremely low energy density. It is almost two orders of magnitudes lower than the photovoltaic electricity. New processes such as ethanol from cellulose are not going to make much difference to this fact. Biomass involves the conversion of solar energy to chemical energy stored in plants and other organic matter. The efficiency of this energy conversion is extremely low. Solar insolation in India is about 1,825 $kWh/m^2/year$, while the net energy density of Jatropha biodiesel is just 3.5 $kWh/m^2/year$! In other words, the energy efficiency is only 0.19%, which results in the low energy density.

The impact of this low energy density on land requirement for biodiesel production is easy to calculate. India consumed about 162 million tonnes of crude oil in the year 2011 [BP 2012]. To replace it with biodiesel will require 540,000 square kilometers of land for Jatropha plantation. Compare this with the land use of 280,000 square kilometers for wheat production in India [UN Data]. This calculation is for Jatropha biodiesel. Ethanol from corn or sugarcane has an even lower net energy density and consequently needs even more land.

It is this land (and water) requirement of biofuels that has been the genesis of the food vs fuel (or drink vs drive!) debate. In the case of India, economic growth has improved the standard of living and the crude oil consumption is rising. At the same time, urban sprawl and industrial production are making increasing demands on land and water resources. Land under agriculture is not increasing, despite the rising population and the increasing demand for food grains [UN Data]. In this situation, finding arable land and water for a large scale production of biofuels is simply out of question. What is true for India is also true for most other developing nations. Secondly, no one knows what impact a large scale plantation of biofuels would have on biodiversity, soil quality, and water availability.

For most countries in the world, forests are already a dwindling resource. Using such forests on a large scale as an alternative source of energy is not a practicable solution. Growing new forests is not an option either, due to land availability and the energy input required for such an activity.

Certain conclusions can be drawn from this analysis:

- For the hundreds of millions of poor people in the world, biomass will continue to be a useful resource. Most of the biomass usage will be in the form of gathered firewood, crop residues, charcoal, and animal dung.
- Biofuels will not be able to replace crude oil. However, they will be an alternative on a smaller scale, available for use near the point of production. Biofuels will be a good alternative to diesel or gasoline in some rural areas.

Creating Concentrated Sources of Energy

Modern civilization uses concentrated sources of energy on a very large scale. Gasoline, diesel, natural gas, and coal are all concentrated sources of energy. Some are either used directly or after conversion into electricity. The electric grid also serves as a media to access and concentrate the sources of energy. The renewable sources of energy are characterized by very low energy density. Table 5.16 lists the estimated energy densities of various renewable sources of energy from the point of view of land-use.

Energy Source	Energy Density (kWh/m²/year)
Hydroelectricity	3 to 80
Wind Energy	2.8
Photovoltaic Electricity	100 to 300
Solar Thermal Electricity	62
Jatropha Biodiesel	3.5

Table 5.16: Energy Densities of Renewable Sources of Energy

The low energy density implies that the resources are dispersed over a wide region. Solar insolation, which is the primary energy source for these renewable sources of energy, is widely dispersed. After applying the energy efficiencies of conversion processes, the resources so created are inherently dispersed. In each case, the renewable energy source has to be concentrated in some form to be of any real use. As a result, obtaining concentrated sources of energy from such dispersed resources will require an energy input in some form. Conversion to electricity and connection to an electric grid is one of the ways of achieving concentration. In the case of hydroelectric power, concentration is achieved by building a dam that creates a large reservoir. The water from the reservoir can be used to generate a huge amount of electric power. In the case of solar photovoltaic electricity, concentration is achieved by connecting a large number of solar panels to the electric grid, over a wide region. In the case of wind energy, some amount of concentration is achieved by the use of very large wind turbines and by connecting many such turbines, distributed over a large area, to the electric grid. In the case of solar thermal electricity, sunlight is concentrated optically by thousands of heliostats onto a central collecting tower. Multiple such systems are connected to the electric grid. For biodiesel, it is necessary to collect the fuel produced in a large geographical region to create a useful supply. It can be observed that there will be costs and energy inputs that correspond to each method of concentration. The challenge in engineering renewable sources of energy will always be to improve energy density and creating concentrated source of energy, while keeping costs and energy inputs low. The improvements will be limited by the energy density of solar insolation and the inherent efficiencies of energy conversions.

6. NUCLEAR ENERGY

Nuclear energy is the only form of energy in which mass gets converted into energy. Albert Einstein formulated the famous equation to calculate the amount of energy generated (E) when mass (m) gets converted into energy:

$$E = m * c^2$$

Here, c is the speed of light in vacuum, which is 3×10^8 meters per second. Thus, the conversion of a mass of 1 kg into energy will yield:

$$E = 1 * (3 * 10^8)^2 = 9 * 10^{16} \ Joules$$

Or, about 90,000 trillion Joules! This is roughly equivalent to the energy obtained by burning 3 billion kg of coal! This calculation is for the purpose of illustration only. In practice, the amount of mass that gets converted into energy in a nuclear reactor is very small. On fission of an atom of an element such as uranium-235 (U-235), a tiny amount of mass gets converted into energy. The energy is generated primarily in the form of kinetic energy of fission fragments. In a nuclear reactor, this kinetic energy gets converted into heat, which is used to generate steam from water. It is the steam that drives turbines to generate electricity.

Nuclear Fission

The conversion of mass into energy in a nuclear reactor requires a nuclear fission reaction to take place. Nuclear fission means splitting the nucleus of an atom of a

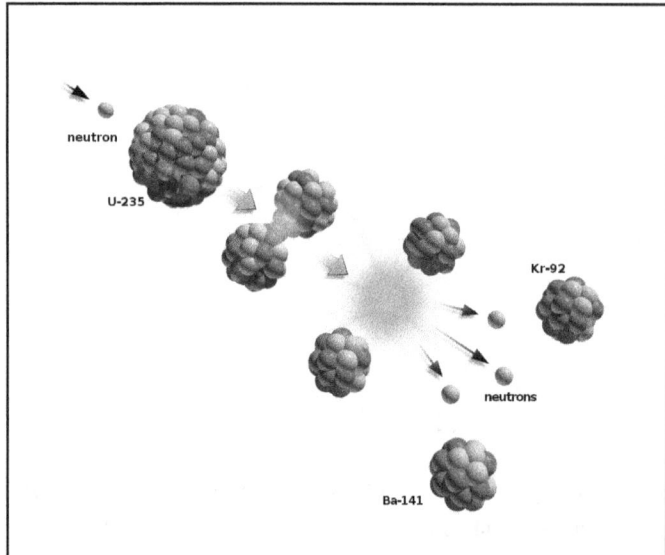

Illustration 6.1: Fission of a Uranium Atom. Credits: Andrea Danti/Shutterstock.com

heavy element such as uranium. This results in the creation of two or more atoms of lighter elements, and at the same time a very small amount of mass gets converted into energy. The fission reaction is typically caused when a free neutron collides with the nucleus of an atom of a fissile element such as uranium-235. The number 235 represents the isotope of uranium that has 143 neutrons and 92 protons in the nucleus. This fission reaction itself yields neutrons, which, in turn, may cause other fission reactions. The fission reaction has to sustain on its own in a nuclear reactor for continuous generation of energy. At the same time, the reaction should not go out of control, which would result in a meltdown!

Fissile and Fertile Materials

Fissile materials are those which can produce a sustained nuclear fission reaction. A sustained fission reaction leads to a continuous generation of energy, which is used to generate electricity in a nuclear power station. There are only three fissile materials of interest for the purpose of nuclear energy: U-235, U-233, and Pu-239. Out of these, only U-235 is found in nature. Pu-239 is an isotope of plutonium which is produced from U-238 after a neutron absorption and two beta decays. U-233 is an isotope of uranium that is not found in nature. It is obtained from thorium (Th-232) after a neutron absorption and two beta decays. Since U-238 and Th-232 can be used to produce fissile material, they are termed as fertile materials. It is the fissile materials that undergo fission in a nuclear reactor to produce energy. Though U-235 is found in nature, its concentration is only about 0.7%. More than 99% of uranium found in nature consists of isotope U-238.

Enriched Uranium

Since only 0.7% of the naturally found uranium is fissile (U-235), it is generally necessary to increase the percentage of U-235 by artificial means, before it can be used as a fuel in a nuclear reactor. This process is known as the process of enrichment. The percentage of U-235 is typically increased to around 4% as a part of the enrichment process. This concentration of U-235 is sufficient for sustaining a chain reaction in a Light Water Reactor (LWR).

Note: There are PHWR reactors that can use natural uranium as fuel. These reactors use heavy water as the moderator. Details on this are beyond the scope and interest, as far as this book is concerned.

Uranium Energy Cycle

As U-235 undergoes fission in a nuclear reactor, some of the released neutrons are absorbed by fertile U-238, converting it into fissile Pu-239. Some of the Pu-239 also undergoes fission in the reactor, producing energy. It is estimated that about 30% of the energy in a U-235 based reactor may actually be obtained due to fission of Pu-239, produced within the reactor. Spent fuel in a reactor contains useful Pu-239 which can be separated with fuel reprocessing technology. This Pu-239 is mixed with U-238 and used as fuel in a nuclear reactor. Again, some of the U-238 in such a reactor will be converted into Pu-239. In theory, it is possible to continue this cycle of conversion of fertile U-238 to fissile Pu-239 until all U-238 is converted.

Thorium Energy Cycle

Thorium is found in relative abundance as compared to uranium in many countries. Thorium is not a fissile element and hence cannot be used directly in a nuclear reactor. However, thorium is a fertile element. It is possible to convert thorium into fissile U-233 with a neutron absorption reaction. Since U-233 is fissile, it is possible to use it as a reactor fuel. In theory, neutrons released in the fission of U-233 can convert more thorium into U-233 in a nuclear reactor. The U-233 formed in this process can be separated and used as reactor fuel. This is a sustainable process that can convert the entire thorium resource to U-233 for the production of energy. The potential of thorium is much larger than that of uranium, thanks to a larger resource base. According to one estimate, if India were to use the uranium energy cycle to produce electricity for meeting the entire projected demand in the year 2052, the uranium resources in the country would last for only 18 years. For the same level of production, the thorium resources in the country would last for almost two centuries [Kakodkar 2009].

Energy Generation Capacity of 1 kg of Uranium

Fission of a single atom of U-235 yields about 200 million electron-volt (Mev) of energy. An electron-volt corresponds to about 1.6 x 10^{-19} Joules. Thus, fission of a single U-235 atom yields about 3.2 x 10^{-11} Joules of energy. A kg of U-235 contains about 25.6 * 10^{23} atoms, based on Avogadro's number. Fission of all U-235 atoms will release about 82 * 10^{12} Joules of energy. This is equivalent to about 22.8 million kWh, which is the same as 22.8 GWh.

However, uranium found in nature only contains about 0.72% of U-235. This translates into an energy generation capacity of 0.16 GWh per kg of natural uranium. Most nuclear reactors use enriched uranium with about 4% U-235 content. Energy generation capacity of such a reactor fuel is calculated to be about 0.9 GWh per kg. In practice, about 30% more energy is generated as some of the U-238 in the fuel converts into Pu-239 which undergoes fission. Thus, a kg of typical reactor fuel will generate around 1.2 GWh of thermal energy. The electrical output can be calculated by using 38% as the typical thermal efficiency. Hence, one kg of 4% enriched uranium can be expected to yield about 0.46 GWh of electrical energy.

Yearly Nuclear Fuel Requirement of a 1 GWe nuclear power station

One GWe is the measure of electrical power output of a nuclear reactor. During a year of non-stop operation at full capacity, such a power station will produce 8,760 GWh of electrical energy. Based on the calculations from the earlier section, such a power station will use about 19 tonnes of enriched uranium (4%) per year. Note that this translates to around 106 tonnes of natural uranium, assuming no losses in various conversion processes from uranium ore to enriched uranium fuel. This calculation also assumes that the entire mass of U-235 fissions in the reactor. In reality, there are various losses in the entire process from mining to electricity generation and not all of U-235 undergoes fission in the reactor. World Nuclear Association reports that the 375 GWe of nuclear reactor capacity in the world requires an annual supply of 68,000 tonnes of uranium. This implies that per GWe capacity, a supply of about 180 tonnes of uranium is necessary every year [WNA 2013a].

Uranium Reserves in the World

Uranium is quite ubiquitous in nature. While deposits containing high concentration of uranium ore are not common, it is found in low concentration widely across the earth. Interestingly, uranium is found in extremely low concentration in granite and even in seawater. Generally, higher the concentration of uranium ore in a deposit, the cheaper it is to extract it. Deposits containing uranium concentration of 2% or more are not common. The amount of uranium classified as a "reserve" depends upon the cost associated with mining it. As discussed in the previous paragraph, a 1 GWe power station requires about 180 tonnes of natural uranium. If the concentration of uranium in a deposit is 0.1%, which is quite typical, mining 180 tonnes of uranium ore will require moving at least 180,000 tonnes of earth. Quite obviously, this task requires energy that is provided by fossil fuel products. As the price of oil varies, so does the cost of uranium mining. For a price tag of USD 130/kg in the year 2009, uranium reserves in some countries are listed in Table 6.1. Uranium reserves are a function of the price tag or the market price of uranium. A higher market price will result in the classification of additional deposits as reserves.

Country	Uranium Reserves (tonnes)
Australia	1,673,000
Kazakhstan	651,000
Canada	485,000
Russia	480,000
South Africa	295,000
Namibia	284,000
Brazil	279,000
Niger	272,000
USA	207,000
China	171,000
Jordan	112,000
Uzbekistan	111,000
Ukraine	105,000
India	80,000

Table 6.1: Uranium Reserves by Country. Source: World Nuclear Association [WNA 2010]

How Much Electricity is Generated by Nuclear Power Plants?

Table 6.2 summarizes the electrical power generation capacity of nuclear power plants. Only the top 14 countries are listed here. This data is obtained from the United Nations database [UN Data]. This data is somewhat dated, as recorded in the year 2009. However, capacity addition since 2009 has not been very significant.

Country	Nuclear Power Capacity (GWe)
United States	101
France incl. Monaco	63
Japan	48
Russian Federation	23
Germany	20
Republic of Korea	17
Ukraine	13
Canada	13
United Kingdom	10
Sweden	9
China	8
Spain	7
Belgium	5
India	4

Table 6.2: Nuclear Power Generation Capacity (Source: UN Data)

How Many Years will the Uranium Reserves Last?

Total installed capacity of nuclear power stations is about 375 GW in the year 2012, and the total uranium reserves in the world are about 5 million tonnes, assuming a cost of $130/kg of uranium. A power station with a capacity of 1 GWe requires around 180 tonnes of uranium per year. A simple calculation indicates that the uranium reserves should last for 75 years, based on existing consumption. This calculation is simplistic, however. It ignores the possibility of fuel reprocessing to retrieve the fissile Pu-239, which can be used as reactor fuel. Secondly, the possibility of the use of breeder reactors has not been considered either. Nevertheless, this calculation is useful to indicate that there is no impending doom as far as the availability of nuclear power is concerned. However, it will be wrong to conclude that nuclear power is the answer to the impending energy crisis, fossil fuel depletion, or global warming, just on the basis of such a calculation.

Is Nuclear Power the Answer for the Energy Crisis?

Due to the rising prices and depletion of fossil fuels, and perhaps more importantly, due to the increase in CO_2 concentration in the atmosphere, there is an increasing pressure to reduce the consumption of fossil fuels. It is suggested that nuclear power is a suitable alternative to fossil fuels. Nuclear power can substitute coal and natural gas based power stations, while being competitive in costs. It is also claimed that nuclear energy has no associated CO_2 emissions. Secondly, if nuclear electricity is inexpensive, it is certainly possible to visualize a number of applications which will make it possible to replace the use of fossil fuels in the transportation sector as well as in heating applications. However, in order to make a global impact on the fossil fuel usage, the production of nuclear power should increase significantly. It is important to take a look at some hard data to get an idea of the scale of such an effort:

The worldwide share of nuclear power in electricity generation was only about 15% in the year 2009 [UN Data]. Note that this was only the share of electricity generation and not of primary energy consumption. In India and China, only two percent of the electricity production is sourced from nuclear power plants. Economies of both these countries are growing rapidly and the energy demand is increasing in synchronization with economic growth. China increased the consumption of coal rapidly in the last 15 years. India is likely to do the same in the next 15 years. Such an expansion will significantly increase fossil fuel consumption and consequently, CO_2 emissions. India and China will need to make use of nuclear energy to satisfy a significant portion of the new demand. Such a usage of nuclear energy will require a rapid expansion of nuclear power capacity in these two countries.

In the year 2010, the TPE consumption in the world was 12,002 MTOE. Nuclear energy supplied only 626 MTOE, or about 5% of the TPE consumption, while fossil fuels supplied 86% of the TPE consumption [BP 2011]. Let the target for growth in the supply of nuclear energy over the next two decades be to supply 30% of the world TPE consumption. The world TPE consumption is assumed to grow at a rate of 3% per year. It can be calculated that in two decades, the TPE consumption will grow to 21,677 MTOE. In order to supply 30% of this TPE consumption, it is necessary to generate nuclear energy equivalent to 6503 MTOE. This calculation implies that the supply of nuclear energy must grow by an order of magnitude over the next two decades. Even with such an increase in production, fossil fuel replacement would not be achieved. In fact, fossil fuel usage will grow unless the renewable sources of energy increase their contribution by more than an order of magnitude. Thus, much more than an order of magnitude increase in the nuclear generation capacity is necessary over the next two decades in order to reduce the usage of fossil fuels. However, such an increase in capacity will deplete the known uranium resources in less than 7 years, assuming no reprocessing of spent fuel or the use of breeder reactors. Realistically, the price of uranium will rise due to an increase in demand, triggering exploration of uranium resources. It is very difficult to predict if the economically extractable uranium resource in earth's crust will increase by an order of magnitude in response to the rising price. However, the rise in demand for fissile material will certainly renew the interest in breeder reactors and in reprocessing of spent fuel.

Nuclear Waste and Reprocessing

Nuclear waste is classified as high level waste and low level waste. High level waste is primarily the spent fuel from a nuclear reactor. This section will only deal with the spent fuel and the usefulness of reprocessing.

Not all fissile material undergoes fission in the reactor core. Fuel rods are removed from the reactor core once their capacity to generate energy is significantly reduced. The spent fuel is very hot as well as extremely radioactive. A PWR reactor that uses U-235 as the fissile material typically has the following composition for the fuel input: 4% U-235 and 96% U-238. Table 6.3 Specifies the composition of spent fuel generated by such a reactor [Makhijani 2010].

Element	Approximate Composition
U-235	0.68%
U-236	0.52%
U-238	93.05%
Fission Products	4.62%
Pu-239 and other Plutonium isotopes	0.99%
Other Transuranic Elements	0.095%
Other Uranium Isotopes	0.02%

Table 6.3: Composition of Spent Fuel from a PWR Reactor

Out of these waste products, Pu-239, being fissile, is the useful element for the purpose of energy generation. U-235 can potentially be useful, provided that it can be separated from other isotopes of uranium present in the spent fuel.

Is It Useful to Reprocess the Spent Fuel?

The proponents of nuclear power claim that the reprocessing of spent fuel not only reduces the problem of nuclear waste management significantly, but also suggest that 95% of the spent fuel is recycled [Kakodkar 2011]. This argument is based on the fact that fission products and transuranic waste are only 5% by weight of the spent fuel. The remaining 95% corresponds to uranium and plutonium, which could be utilized in a nuclear reactor. In other words, recycling 95% of spent fuel implies that the utilization of U-238 improves as a result of processing the spent fuel. It is worth analyzing these claims.

A reprocessing plant separates nuclear waste into three streams:

- Plutonium
- Uranium
- Fission Products and Transuranic Elements

The plutonium stream is mostly composed of fissile Pu-239, along with the non-fissile isotope Pu-240. The uranium in the spent fuel consists of several isotopes, specified in Table 6.3. The fission products and transuranic wastes separated from the spent fuel are extremely radioactive and dangerous to the biosphere. These must be safeguarded for centuries. The following points deal with some important aspects for each of these streams:

- The uranium stream contains some U-235 which did not undergo fission in the reactor. It is possible to carry out the uranium enrichment process to increase the U-235 composition in order to be suitable for use as reactor fuel. However, the isotopic composition of uranium in the waste is different from that found in the nature. Specifically, there is U-236, which is detrimental to the chain reaction that takes place in the reactor. There are also some other isotopes of uranium in the waste, present in trace quantities. It is not possible to separate these isotopes using chemical processes and very expensive to separate them with other processes. These isotopes contribute to the difficulties in making use of uranium separated from waste. The isotopic composition degrades with every round of fuel reprocessing, making this task progressively difficult and expensive [Makhijani 2010].

• The amount of plutonium that is recovered from spent fuel is only about 1% by weight. The recovered plutonium is classified as reactor-grade since it contains the non-fissile Pu-240 in significant quantity, though fissile Pu-239 is the primary constituent. Reprocessing of spent fuel thus yields fissile material, useful as reactor fuel. However, it will be important to calculate the amount of such material generated.

• It can be calculated that for one tonne fuel input to a PWR with 4% U-235, it is necessary to start with about 6 tonnes of natural uranium, which contains 43 kg of U-235 and 5,957 kg of U-238. After the process of enrichment, a tonne of fuel input to the reactor will contain around 40 kg of U-235 and 960 kg of U-238. This implies that almost 5 tonnes of U-238 remains as depleted uranium. The spent fuel from such a reactor contains about 10 kg of plutonium, 6.8 kg of U-235, and about 930 kg of U-238, per tonne of fuel input. This data implies that out of the original input of 5,957 kg of U-238, only 30 kg transformed due to neutron absorption in the reactor to plutonium. Some of the Pu-239 gets fissioned in the reactor, while around 10 kg of Pu-239 and Pu-240 remains as a part of spent fuel. Reprocessing of spent fuel will make it possible to use this plutonium. Conversion of 30 kg of U-238 out of an initial quantity of 5,957 kg implies that the utilization of U-238 is only 0.5% after one cycle of reprocessing. Secondly, reprocessing yields only about 25% fissile material as compared to the initial fuel input. Thus, every subsequent reprocessing cycle rapidly reduces the amount of available fissile material. As a result, the energy generation capacity of uranium available in nature is improved only marginally with fuel reprocessing, and most of the U-238 will wind up as depleted uranium. The only way to improve the utilization of U-238 is by using a large number of fast breeder reactors which will be covered in the next section.

• The third stream from reprocessing consists of fission products and transuranic elements. While these constitute only about 5% by weight, the fission products are intensely radioactive and remain a major hazard for at least a few centuries. There is also a significant amount heat production as the fission products undergo radioactive decay which complicates their storage. Another hazardous material constituting the third stream are the so-called transuranic elements. These are the elements heavier than uranium, formed in the reactor as a result of neutron absorption. Half-life of these elements is in thousands of years, and they remain a hazard to the biosphere that requires storage in deep geological repositories. Even the fission products need to be stored in such repositories due to the risk to biosphere for a very long time. As a part of fuel reprocessing, fission products and transuranic waste are dissolved in solvents which are later solidified using a process called vitrification. As a result of this process, the volume of the material that needs to be stored actually increases as compared to the volume before reprocessing. Secondly, reprocessing itself generates low-level nuclear waste that needs to be disposed off securely. Thus, reprocessing does not reduce radioactivity, while increasing the volume of material to be stored [Makhijani 2010].

In conclusion, the only advantage of reprocessing of spent fuel is the extraction of plutonium, which can be used as fissile material in reactors and possibly in nuclear weapons. However, the quantity of plutonium produced is too small to improve the utilization of U-238 significantly.

Breeder Reactors

In a nuclear reactor it is possible to convert fertile elements into fissile elements. For instance, fissile Pu-239 is produced from fertile U-238 after absorption of a neutron, and fissile U-233 is produced similarly from fertile Th-232. This gives rise to the possibility of *breeding* more fuel than what is used up in the fission reaction. Such a reactor can, at least in theory, generate energy as well as breed more fissile material than it uses! It is this idea that gave rise to the possibility of generating energy in such a large quantity, that in the 1950s and 60s there were claims that energy would soon be too cheap to meter! Such reactors could have solved the problem of energy availability for centuries. In practice, breeder reactors have not become commercially or technologically successful, though more than hundred billion dollars have been spent in their research and development over six decades.

Fast Breeder Reactors

A *Fast* Breeder Reactor (FBR) uses *fast* neutrons for the fission reaction. This implies that neutron energy is not moderated in such a reactor. Conventional nuclear reactors use water as the coolant which also acts as a moderator for neutrons. In such reactors, a little over 2 neutrons are produced on the average, on fission of a U-235 atom. Out of these, at least one neutron is necessary for the continuation of the fission reaction. To breed more fissile material than what is consumed, at least one neutron should be absorbed by the fertile material for conversion into fissile material. In practice, there are neutron losses (due to unwanted absorption/escape) in the reactor as well as burn-up of fissile material produced in the reactor. This results in the generation of less fissile material than what is consumed.

In a fast reactor, liquid sodium is used as the coolant. Use of liquid sodium makes sure that neutron speed is not moderated to *thermal* (implies lower energy) range. Fission with fast neutrons can yield a little over 3 neutrons per fission when Pu-239 is used as the fissile material. This improved neutron availability makes it possible to convert more of fertile material into fissile material. Pu-239 also has a better probability of fission with fast neutrons as compared to U-235, making it easier to sustain chain reaction to maintain criticality. Generally, FBRs use a *blanket* of fertile material around the core. The blanket can either contain U-238 in the form of depleted uranium, or thorium. The core contains Pu-239 as the fissile material, mixed with U-238. Part of the fertile material in the blanket gets converted into fissile material as a result of neutron absorption. The blanket is chemically processed to separate the fissile material. Such a reactor, at least in theory, can breed more fissile material than what is consumed. For FBRs, after the initial load of fissile material, no external supply of fissile material should be necessary. All the fissile material is generated within the reactor. At least a few countries have a large stockpile of Pu-239 which can be used to start a large number of breeder reactors. Yet, the number of such reactors currently in operation are only 3 [WNA 2012]. One advantage of such reactors is the absence of chain reaction in the blanket. This makes it much easier to process the blanket and the fissile material obtained is much pure in isotope composition as compared to that obtained by reprocessing the spent fuel. In other words, FBRs can produce weapons-grade fissile material, a proliferation risk from the point of view of nuclear watchdogs.

Breeding Ratio

Breeding ratio is the ratio of fissile material generated to the fissile material used in a breeder reactor. This ratio should be greater than 1 for the reactor to operate as a breeder. For instance, if a reactor generates 5% more fissile material than it consumes, the breeding ratio will be 1.05. The excess fuel so generated may be used to start new breeder reactors, allowing for expansion of capacity. For instance, in the case of thorium fuel cycle, U-233 is used as the fissile material. U-233 is not found in nature and must be bred from thorium. To achieve a large installed capacity of reactors based on thorium fuel cycle, it is necessary that a correspondingly large quantity of U-233 is available. However, the availability of U-233 itself depends on its production in the breeder reactors. This is where the breeding ratio plays an important role. With a breeding ratio of 1.05, it will be possible to double the available fissile material in about 14 years, allowing for doubling of the installed capacity. Breeding ratio is similarly important for the plutonium based fast breeder reactors. A fast breeder reactor produces plutonium much more efficiently as compared to a burner reactor such as a PHWR. To operate a large number of such reactors, it is necessary to obtain a corresponding initial supply of Pu-239. Unless there is an existing stockpile of Pu-239, possibly from decommissioned nuclear weapons, the best way to generate a fresh supply is to use the fast breeder reactors themselves. However, in such a case, the breeding ratio and the corresponding doubling time are of great importance. Increasing the generating capacity from 1 GW to 100 GW will take about 95 years for an FBR technology with a breeding ratio of 1.05. A breeding ratio of 1.1 will require 48 years while a breading ratio of 1.01 will necessitate 468 years for this capacity expansion. The time factor is very important when a country banks its energy future on nuclear energy.

Fast Breeder Reactors: Operational Problems

The concept of breeder reactors has been around for more than 60 years and there were dreams of breeder reactors being the ultimate energy source. There was a projection made in 1976 that 540 commercial FBRs will be operational in the world in the year 2000 and that this number would grow to 2766 by the year 2025 [Cochran 2010]. However, in the year 2010, BN-600 in Russia is the only reactor that is operating and is connected to the grid.

The first commercial FBR to start operations was the Fermi-1 at Lagoona Beach, near Detroit, in USA. It had a power rating of 66 MWe. Fermi-1 achieved criticality in 1963. This reactor was shutdown for almost four years after suffering a partial meltdown in 1966 due to a blockage in the flow of coolant to a fuel assembly. John G. Fuller wrote a book on this accident titled, *We Almost Lost Detroit*. Even after a restart in 1970, Fermi-1 could only operate at a dismal capacity factor of 3.4% in the year 1971. This reactor was finally shutdown in the year 1972. The largest commercial breeder reactor to be built was the SuperPhénix in France, generating 1,200 MW of electricity. This reactor achieved criticality in 1985 and grid connection in 1986. There was a sodium leak in 1987 that led to a shutdown for almost two years. Even after a restart, there were recurring technical difficulties that led to final shutdown in the year 1996. During its lifetime, SuperPhénix achieved only a 7% capacity factor. Japan built a FBR at Monju with a capacity of 280 MWe. There was a sodium fire at this reactor in 1995 that led to its shutdown for major repairs [Cochran 2010]. This reactor is still not operational.

India is building a 500 MWe Prototype FBR at Kalpakkam, near Chennai, in the state of Tamilnadu. This reactor was expected to be ready in 2010. However, there have been delays and it is now expected to be commissioned in 2013. This reactor had received its initial approval in 1990. The reactor is designed based on the experience gained from the operations of the Fast Breeder Test Reactor (FBTR) at Kalpakkam. This reactor had its set of problems which led to several shutdowns. One of the accidents caused a shutdown for two years, while another involved a radioactive sodium leak [Ramana 2009].

The Russian FBR program was also plagued by operational problems. However, the BN-600 remains to be the only FBR still connected to the grid. It has operated with a reasonable capacity factor of 73% since its grid connection in 1980, in spite of 27 sodium leaks and 14 fires until 1997 [Ramana 2009, Cochran 2010]. However, in the last 32 years, no new FBR was connected to the grid in Russia.

A common factor in the breeder reactor design has been the use of liquid sodium as the coolant. At the temperature at which the reactor operates, sodium is highly reactive with water as well as air. Any leak in the pumping system that gets sodium in contact with water or air can lead to an accident. This not only makes FBRs quite expensive to build but an accident typically requires complex cleanup operations. The operational problems and accidents lead to long down-times and repair costs. Another important point is that the reactivity of the FBR core increases in the event of a loss of coolant accident. This constitutes a positive feedback that could lead to an uncontrolled heating of the core and hence a meltdown. There is also a possibility that a meltdown of core can sufficiently disturb the core geometry, causing excessive amounts of fissile material to come together. Such an event could cause a catastrophic accident.

India, Russia, and possibly China, are the only countries that have included FBRs as a part of long term nuclear energy strategy. So far, the main objective of the breeder design, continuous increase in the supply of fissile materials allowing more reactors to be built, is far from being fulfilled. At the same time, there are doubts about safety, cost, and even feasibility of FBRs.

India's Three-Stage Nuclear Energy Program

India achieved a major milestone in the year 2009 by signing the nuclear deal with USA and later with France and Russia. This deal effectively ended the pariah status of the Indian atomic energy program. While the first nuclear reactor to be built in Asia was in India, the installed capacity of nuclear power plants in India was only about 4.7 GW in the year 2010. These power plants have been operating at a very poor capacity factor, estimated to be less than 40%. Normally, the nuclear power stations are expected to operate at a capacity factor exceeding 80%. One of the reasons for such a poor capacity factor is the availability of uranium. India is not blessed with very productive domestic sources of uranium. It was impossible for India to procure uranium from international suppliers since India is not a signatory to the Non Proliferation Treaty (NPT). International sanctions were also in place due to the nuclear weapons tests carried out by India in the year 1998. While the nuclear deal has cleared the way for the supply of uranium and for building new power stations, the long term goal of the Indian nuclear power program is to generate energy from thorium using indigenous technology.

India does not have very large deposits of natural gas, which is a very clean fuel for the generation of electricity. Neither are the deposits of uranium very significant or of high quality. At the same time, the demand for electricity has been growing exponentially with economic growth and population growth. Meeting this demand with coal based generation is a problem, and India has been importing significant amount coal since the year 2010 (See chapter 4.). According to certain estimates, the domestic sources of coal will last for only 12 years if the entire projected demand for electricity in the year 2052 were to be satisfied using coal. This number for uranium is only 4 months! However, India has very large deposits of thorium which can be used to breed fissile material. The domestic deposits of thorium will last for more than 170 years while supplying the entire demand for electricity in the year 2052 [Kakodkar 2010]. At the current rate of electricity production in the year 2010, the thorium resource will last for about 1,500 years. Quite obviously, India has a lot of incentive for pursuing thorium as a source of energy. While the rest of the world has not made much progress in using the thorium fuel-cycle, India has developed a complete plan and designed reactors for the use of thorium. This program was envisioned by Dr. Homi Bhabha almost fifty years ago. It is a very ambitious long-term program, divided in three stages. Each stage involves reprocessing of spent fuel to extract fissile material:

Stage 1: Building uranium based, Pressurized Heavy Water Reactors (PHWR). While these power stations generate electricity, they also generate fissile plutonium (Pu-239). The spent fuel will be processed to extract Pu-239. It is planned to build a PHWR capacity of 10,000 MWe by the year 2020.

Stage 2: This stage consists of plutonium based fast breeder reactors. These reactors will either use a thorium blanket to produce fissile U-233, or a uranium blanket to mutate U-238 to produce fissile Pu-239. The fast breeder reactors will not only breed Pu-239 to start more such reactors but will also generate U-233 for the third stage of the program.

Stage 3: The U-233 produced in the second stage will be used as the fissile material for the thermal breeder reactors in the stage-3. These reactors will mutate Th-232 to breed U-233, while generating electricity. This process will be self sustaining.

It is planned to reach a nuclear generation capacity of 273 GWe in the year 2052, once the three stage program is implemented. This number is more than double the nuclear power generating capacity in the US. In fact, it is possible to reach even higher generation capacity by the year 2052 by increasing the initial PHWR capacity to 40,000 MW in the year 2020 [Kakodkar 2010]. This is an extremely ambitious plan which will achieve energy independence for India. Quite obviously, there are numerous hurdles that range from nuclear physics to economics, and from ecology and safety to global politics. That perhaps is the topic of another book. For the purpose of this book, the next paragraph addresses the time period of several decades that this program requires.

Limits on Generation Capacity and Growth Rate

Almost all the fissile material that is used in nuclear reactors originates from U-235 that exists in nature. There are three fissile materials capable of use in nuclear reactors: U-235, Pu-239, and U-233. Out of these, only U-235 is found in nature in small quantities, while U-233 and Pu-239 have to be bred from Th-232 and U-238, respectively. To achieve a significant increase in the installed capacity of nuclear reactors, it is necessary to secure a supply of fissile materials. For instance, if India were to build nuclear reactors to supply the entire demand for electricity in the year 2004 with nuclear power, the domestic uranium resources would be consumed in only 4 years [Kakodkar 2010]. The limited availability of fissile U-235 restricts any significant expansion of installed capacity of nuclear reactors. The only way out of this problem is to breed fissile material. Breeding of fissile material takes place as a result of neutron absorption. The neutrons produced as a result of fission of U-235 are capable of breeding more fissile material. However, the fission of a U-235 nucleus produces only 2.4 neutrons on the average. Out of these, a little over 1 neutron is consumed on the average for the continuation of fission reaction within the reactor. This leaves only a little over 1 neutron available for the production of fissile material. The PWR and PHWR reactors are not very efficient at using the available neutrons to generate fissile material. As a result, such reactors are net consumers of fissile material. The only viable option for growing fissile material are the fast breeder reactors which are specifically designed to breed more fissile material than what is consumed. The breeding ratio of such reactors determines the time within which the supply of fissile material will double. The first fast neutron reactor was the EBR-1 in the US in the 1950s. In spite of more than 60 years of operations, there is no evidence of ability of fast breeder reactors doubling the fissile material in a reliable, safe, and commercially viable manner, in a reasonable amount of time. As a result, very few such reactors are in operation and that limits the supply of fissile material to embark upon a massive plan of capacity expansion. The only way to solve this problem is to look for some other source of neutrons such as those obtained with the help of a particle accelerator. Such neutrons can also be used to irradiate fertile material to generate fissile material. However, such systems are far from being commercial, any time soon. These are some of the reasons that the three-stage nuclear energy program proposed by India will take a long time to implement, even after the initial PHWR capacity is online by the year 2020. The performance and safety of fast breeder reactors will be extremely critical to the success of this program. It must be questioned if the decision to bank the energy future of the country on a technology that is inherently unsafe and unproven is wise one, given the experience of past 5 decades.

Energy Density and EROEI for Nuclear Energy

It was discussed in an earlier section that fission of 1 kg of U-235 will yield approximately 82,000 GJ of thermal energy. However, 1 kg uranium obtained in nature contains only 0.72% U-235. As a result, one kg of natural uranium will yield about 574 GJ of thermal energy on fission of the contained U-235. This is about 12,500 times the energy density of gasoline. It is theoretically possible to convert the entire quantity of U-238 to Pu-239 in a breeder reactor. Taking the thermal energy released on fission of Pu-239 into account, the energy density of uranium works out to be around 82,000 GJ/kg or 1.75 million times the energy density of gasoline.

It is this phenomenal energy density which makes the calculations of EROEI and EUF redundant for nuclear energy. The only case when such calculations could be useful is when uranium is harnessed from granite or seawater, where uranium is contained in very low concentration, and when breeder reactors are not implemented to fully utilize U-238. However, uranium resources on the planet won't be exhausted for a long time at the current rate of consumption and there may never be a need to harness uranium from seawater.

Nuclear Energy: A Green Source of Energy?

The ecology and environment related aspects of nuclear energy will not be discussed much in this book. These are complex topics that require a detailed analysis of various aspects. An earlier section briefly discussed nuclear waste and reprocessing. This section will only examine the claim by the proponents of nuclear power that there are no CO_2 emissions for nuclear power [Kakodkar 2011].

There are three major stages in the life cycle of a nuclear power plant:

- Reactor Construction
- Power Generation
- Decommissioning

The construction cost of a nuclear reactor (PWR) is generally assumed to be around $2000 per kWe capacity. For instance, the Areva nuclear power project in Jaitapur, India, is expected to cost about $20 billion for the 9900 MWe plant [Chopra 2011]. The activity of reactor construction is generally expected to take about 5 to 7 years. However, there is a history of time and cost overruns for many projects. For instance, the Areva reactor in Finland was supposed to have entered service in the year 2009. However, it won't be producing electricity until 2014. There is also a cost escalation of €2 billion for the Finland project [Boxell 2012]. A major part of the huge capital expense of any nuclear power plant involves site preparation, manufacturing, and construction costs. These require direct consumption of fossil fuels which results in CO_2 emissions. Another significant part of the cost corresponds to the internal operations of the company that is constructing the reactor and its partners who are part of the business ecosystem. It can be observed that the capital cost of a nuclear reactor represents a sufficiently complex economic system so that the energy intensity of economy can be a useful measure to estimate the energy input that it corresponds to. In a fossil fueled economy, this energy input directly translates to CO_2 emissions. Thus, apart from the direct CO_2 emissions which result from the construction activity, there are indirect emissions as well. Secondly, the cost of a reactor generally does not include the political, governmental, and military overheads that are associated with nuclear power. This would be an indirect subsidy from the government to nuclear power, the cost of which would be very difficult to estimate. However, these indirect costs also correspond to some CO_2 emissions. Thus, even before a nuclear reactor generates any electricity, significant CO_2 emissions do take place that correspond to the direct and indirect costs of a nuclear reactor.

During the power generation phase, CO_2 emissions take place due to uranium mining and processing it into fuel for the reactor. It was calculated earlier that a 1 GWe power station requires around 180 tonnes of uranium per year. With an ore concentration of 0.1%, it is necessary to mine a mass of 180,000 tonnes per year per GWe capacity. This is an order of magnitude less than a coal-based power plant of equivalent capacity. Quite surely, the operations phase of a nuclear reactor is responsible for very low CO_2 emissions.

Emissions also take place as a part of storage and processing activities for nuclear waste. This part is very difficult to quantify. Firstly, a fully operational deep geological repository for nuclear waste has not been constructed yet. Yucca mountain in Nevada, USA, was a proposed site for nuclear waste from commercial reactors in USA. Until such a repository is in place, it is necessary to safeguard the nuclear waste. The time frame for this activity could be very long, possibly more than 1,000 years. A similar argument also applies to decommissioning of a nuclear reactor. While the deconstruction of a nuclear reactor is in itself a complex activity that will involve direct and indirect CO_2 emissions, it is also necessary to safeguard the high level and low level radioactive waste that is generated as a result of decommissioning.

In summary, the argument that nuclear energy is free of CO_2 emissions is simply not true. However, it can certainly be stated that nuclear energy has lower emissions as compared to fossil fuel based power plants, thanks to the low CO_2 emissions during the operational phase of a nuclear reactor.

Nuclear Renaissance and the Fukushima Accident

Nuclear renaissance refers to the hopes of restarting the activity of building new nuclear power plants in the OECD nations. There were two accidents in the last century which virtually stopped the work of building new nuclear reactors in these nations, with the exception of Japan and France. The first accident took place at Three Mile Island in the US, and the second one was the Chernobyl disaster in Ukraine (a part of USSR at that time). Both these accidents polarized public opinion strongly against nuclear power. Governments denied new permits for nuclear power plants and even canceled some projects that already had the permits. Since 1991, only two new nuclear power plants were connected to the grid in the US. Most of the new nuclear reactors connected to the grid since 1991 belonged to China, Russia, India, Japan, France, and South Korea [WNADB]. However, over the last decade, there were some drivers which renewed the hopes of a renaissance for the nuclear industry, particularly in the US:

- Global warming and the associated CO_2 emissions became a concern. Nuclear energy is advocated as an energy source with no or low CO_2 emissions.
- The high price of crude oil and its impact on the economy became a serious concern. Nuclear energy is perceived as a low-cost source of energy.
- The 9/11 attacks on the US gave a new thrust to the need for energy independence in the US.
- With no new catastrophic accident since the Chernobyl disaster in 1986, there was some acceptance of, or at least a lesser resistance to, nuclear power in the public opinion.

However, on March 11, 2011, Japan was hit by a catastrophic earthquake of magnitude 9.0. This was followed by a tsunami that led to an accident at the Fukushima Daichi nuclear power complex. The Fukushima complex consisted of 6 nuclear reactors. Out of these, three units were operational when the earthquake struck, while three units were in a planned shutdown state. On detecting seismic activity, the operational units were shut down automatically. However, the 15 meter tsunami that followed the earthquake inundated the nuclear power complex. It damaged every backup system, leading to a total power blackout. A nuclear reactor must be cooled actively, even when it is in a shutdown state. The decay of fission products was producing about 22 MW of thermal power in reactor 1, and 33 MW for reactors 2 and 3. With an electrical power blackout, there was a failure of the cooling system. As water started to boil, pressure started to build in the reactor vessels. Eventually, fuel rods were exposed, and without a coolant, the temperature of the fuel rods rose in an uncontrolled manner. Finally, this not only led to a core meltdown but also to explosions due to a buildup of hydrogen. The reactors had to be cooled externally by spraying seawater using fire pumps. The fire and water leaking from the reactor complex released a significant amount of radioactive material in the environment. Population in the radius of 20 km was ordered a compulsory evacuation due to the dangers posed by the release of radioactive material. This resulted in a displacement of 160,000 people. The evacuation orders still remain in place in most of this area at the end of the year 2012. Four of the 6 reactors were completely damaged, resulting in a loss of 2.7 GW of generating capacity [WNA 2013]. Even the remaining two reactors at Fukushima could remain in a cold-shutdown state for a very long time.

The catastrophic nature of events at Fukushima will make the task of a nuclear renaissance very difficult. In spite of improvements in the reactor design, it will be an uphill battle to convince people of the safety of nuclear power. Even though the demand for energy is ever increasing, no one wants a nuclear power plant in their backyard. Soon after the Fukushima disaster, Germany announced plans to shut down all nuclear power plants by the year 2022. Japan shut down all 50 nuclear reactors, but might restart at least some of them. While there are no plans to abandon nuclear power in the US, a renaissance certainly looks doubtful. Finally, there is no indication that India will abandon the ambitious nuclear power plans in spite of growing opposition witnessed for the Jaitapur and Kudankulam power projects.

Conclusion

Nuclear energy has always been a controversial source of energy. While it is impossible to predict the future for nuclear energy, it is possible to draw certain conclusions based on the discussion in this chapter:

1. There is no impending doom as far as the supply of uranium is concerned. At the current level of power generation, known reserves of uranium will last for about 75 years. Any shortages of uranium will be localized in certain countries due to limitations of local supply, restrictions on imports due to proliferation concerns, and international politics.

2. Nuclear energy is an important source of electricity is many countries. However, it supplied only about 5% of the TPE consumption in the world [BP 2011]. For nuclear energy to partially replace coal and natural gas in the next two or three decades, a significant expansion is necessary in the nuclear generation capacity. Such an expansion is also necessitated by the rapid growth and industrialization of India and China.

3. An order of magnitude increase in the nuclear generating capacity will rapidly deplete the uranium resources of reasonable quality (cost). There will be a need to use fast breeder reactors to breed plutonium from U-238 and U-233 from thorium.

4. Fast breeder reactor technology continues to be unproven and risky despite six decades of research. Banking the energy future on this technology is risky not

only from the point of view of inherent risks and reactor safety, but also from the point of view of proliferation of weapons-grade fissile material.

5. Problems in FBR implementations and low breeding ratios lead to long doubling time for fissile material. Non-availability of fissile material is certain to delay the plans of building new power plants based on plutonium or thorium fuel-cycle. This time can well be used to harness renewable energy sources, which might alleviate the need for large capacity expansion of nuclear power.

6. The catastrophic nuclear accidents at Fukushima and Chernobyl have etched a fear of nuclear energy on the minds of people. There will always be an opposition to new nuclear power plants. While proponents of nuclear power make claims about improvements in design and safety of nuclear reactors, there will always be a finite probability of occurrence of a catastrophic accident. In a world where wars are still being fought and terrorism is a major concern, it is possible to think of ways in which nuclear reactors can be targets of attacks that can lead to catastrophic accidents. Even the highly radioactive spent fuel can be used to manufacture *dirty* bombs that can have deadly consequences. It is quite pertinent to ask, "Is our civilization mature enough to be a user of nuclear energy?"

7. LOOKING INTO THE FUTURE

This tour of the world of energy resources will be complete after taking a look at the future. The analysis so far indicates that the next couple of decades will be quite interesting. At the end of the year 2012, the developed nations are in a prolonged state of recession, or at least in a prolonged period of economic slowdown. India and China are still growing in population and in economy. It was discussed in chapter 3 that every dollar in circulation is finally a claim on energy. Hence, the economic growth in India, China, and in the other developing nations is leading to an exponentially increasing demand for energy. The economic slowdown since the year 2008 has resulted in a reduction in energy consumption by the OECD nations. However, every nation seeks economic growth. In fact, economic growth is a necessity of the monetary system. It is only a matter of time when the OECD nations return to the path of economic growth and consequently increase their energy consumption. Aspirations are increasing all over the world as nations seek to alleviate poverty. Rising human development brings about a corresponding rise in energy consumption. Finally, population of the world continues to grow. A rising population, improving human development, and economic growth lead to an ever increasing demand for energy.

At the same time, the energy supply situation is very unclear. The international price of crude oil remains over $100 per barrel. This high price has not resulted in a significant increase in supply. The share of TPE consumption supplied by coal has increased in the 2000-10 decade. The high demand for coal has led to an increase in coal prices. It is quite disappointing that the consumption of coal, which is the most polluting of the fossil fuel resources, increased in a decade in which there was a widespread acceptance that global warming is indeed a reality. Natural gas, which is certainly a better fuel, was unable to displace coal. During this period, renewable sources of energy are certainly gaining more acceptance. However, their share of the total energy supply remains very small (not counting hydroelectricity). It will take decades for the renewable sources of energy to increase their share sufficiently to help reduce the consumption of fossil fuels. There is no such thing as an ideal source of renewable energy and each source has some or other limitation. During the 2000–10 decade, there were some hopes of a nuclear renaissance, which were dashed after the disaster at Fukushima. It does not seem likely that the nuclear energy output will rise sufficiently in the next couple of decades to help reduce the consumption of fossil fuels. The non-availability of alternatives to fossil fuels is certainly a problem when there are real concerns about peak-oil being around the corner. High oil prices over the last few years can certainly be an indication that the peak in production for conventional, inexpensive crude oil, has very likely been reached.

An insufficient energy supply implies rising energy prices that are certain to have a negative impact on the economy. Perhaps more importantly, this has a negative impact on human development and poverty alleviation. There are several websites on the internet, dedicated to the topic of peak-oil and its impact on civilization. Some of them go to the extent of predicting an end of civilization due to peak-oil. In this book, no such extreme prophecies will be made. It was discussed in this book that there is no imminent doom due to a sudden stoppage of energy supply. However, it was also discussed that there is a problem in growing the supply of energy resources exponentially from the current level, as demanded by economic growth. Every outstanding dollar in the economy is a claim on energy resources. The corollary of this statement is that a lack of increase in energy supply will lead to a stagnation of economy.

The environmental impact of a limited supply of clean energy could be severe as well. There will be a tendency to make use of coal and heavy/non-conventional oil. Both these sources not only lead to environmental pollution but also to an increase in CO_2 emissions, for a given supply of energy. The science and mathematical models related to global warming are still evolving. However, there is a fear that there may be certain tipping points, which will be reached due to an increasing CO_2 concentration in the atmosphere. A catastrophic and irreversible climate change may take place once the tipping points are breached. So far, the international negotiations to curb the CO_2 emissions have not proven to be successful. In fact, the recessionary pressure since 2008 makes it rather difficult to accept any mandatory cuts in emissions. The integrity of biosphere is obviously more important from the point of view of civilization than any short term targets regarding economic growth.

Clearly, there are challenges ahead for which no easy solutions exist. While there are limitations in increasing the supply of energy, any reduction in energy consumption is certain to have an impact on economy. A reduction in energy consumption in the developing nations will also impact human development in these nations. One obvious solution is to reduce the energy intensity of economy by improving the energy efficiency wherever any energy-use takes place. Such an improvement may be able to achieve the same economic output while using lesser energy. This change will also help achieve the same human development while consuming lesser energy. It is certainly worth analyzing this option.

Impact of Energy Efficiency Improvements

There are experts who believe that an ample amount of energy can be saved, simply by improving the energy efficiencies of all devices. Such savings will lead to a reduction in the demand for energy resources which will help stabilize, or reduce, their prices. Stable or low energy prices are well suited for economic growth. In this section, the topic of energy efficiency improvements and their impact on total energy consumption will be discussed.

In the case of electricity generation, there is still a tremendous amount of scope to improve the efficiency of thermal power plants, simply by replacing the old units with new ones that use modern technology. The transmission and distribution losses for electricity are very high as well, especially in the developing nations. These losses can be reduced by upgrading the electric grid, and the local, last-mile, distribution network. These upgrades are expensive and capital-intensive. Access to capital and the cost of capital is always a problem that results in the use of inefficient equipment and in the continuation of use of the outdated equipment that has poor energy efficiency. These problems are especially prevalent in the developing nations.

In the transportation sector, hybrid cars are becoming more popular. These cars employ regenerative breaking to charge batteries when breaks are applied. In other words, kinetic energy of the vehicle is converted into chemical energy, which is stored in the battery. The energy stored in the battery is used later, to drive the vehicle. Use of hybrid vehicles certainly leads to a reduced consumption of gasoline per kilometer. However, the penetration of this technology is rather limited. The cost of hybrid vehicles tends to be higher than the equivalent, non-hybrid, models. The number of existing vehicles are also too many to replace in a short time period.

National level programs that rate appliances based on energy efficiency are useful. They allow the consumer to make informed choices. In general, the efficiency of all domestic appliances have improved significantly as compared to those a couple of decades ago. This fact is true for all appliances, whether it is television sets, water heaters, washing machines, air conditioners, or refrigerators. The efficiency of computers have increased by orders of magnitude when measured in terms of CPU speed or the speed of data movement.

It will be intriguing to check if the improved efficiencies have resulted in any reduction in the energy use per capita.

Year	Population (million)	TPE Consumption (MTOE)	GDP million $ (2005)	TPE per capita (TOE)	Energy Intensity (TOE/$)
1980	229.8	1,813	5,796,400	7.89	313
1985	241.1	1,767	6,795,600	7.33	260
1990	253.3	1,957	7,962,600	7.73	246
1995	266.3	2,102	9,019,900	7.89	233
2000	282.5	2,314	11,158,100	8.19	207
2005	296.8	2,351	12,564,300	7.92	187
2010	310.4	2,286	12,992,000	7.36	176

Table 7.1: Per Capita TPE Consumption and Energy Intensity Over the Years for USA. Data Source: United Nations for population and GDP numbers; BP for TPE Consumption [UN Data, BP 2011].

Table 7.1 calculates the per capita TPE consumption in the US from 1980 to 2010. This table also calculates the energy intensity of US economy over the same time period. For the purpose of these calculations, US dollars are inflation adjusted to the year 2005. It can be observed that the energy intensity of economy has decreased consistently. In the year 2010, about 44% less energy was required per inflation adjusted US dollar of economic output. While a part of this improvement could be due to a change in the composition of economic output, improvements in energy efficiency of systems and processes must have played a useful role too. Despite a significantly reduced energy intensity, the total energy consumption consistently increased during this time period. While a part of this increase can be attributed to a growing population, it is the growing economy that demanded the increase in energy consumption. Even the per capita energy consumption was slightly higher in the year 2005 as compared to the year 1985, despite the significant improvements in energy efficiency. The per capita energy consumption declined in the year 2010, which was probably due to the lasting impact of economic recession that started in the year 2008.

This data indicates that the hypothesis that improvements in energy efficiency brings down energy consumption is not true at the macro level. There are two reasons for such a counter-intuitive conclusion. Firstly, the efficient appliances and devices are generally expensive than the less-efficient ones. For instance, LED lights are much more expensive as compared to the incandescent and CFL lamps; refrigerators and air conditioners with the highest energy efficiency ratings are expensive as compared to the ones with lower ratings; and hybrid cars are more expensive as compared to the corresponding non-hybrid models. The extra capital cost of high efficiency devices corresponds to business ecosystems that need energy consumption of their own. For instance, the large battery pack, electronics, and control software in a hybrid vehicle correspond to a new business ecosystem which demands a certain energy consumption. Every dollar in circulation is a claim on energy and the additional expense of an energy efficient device corresponds to certain additional consumption of energy. Thus, as the efficient devices replace the less efficient ones, the corresponding economic activity does come at the price of an additional energy consumption. Secondly, though highly efficient devices continue to replace the ones with lower efficiency, requirements increase which result in higher consumption. For instance, there is a tendency to buy bigger refrigerators, larger homes, and drive longer distances, though using highly efficient devices. Thus, any gains due to a better energy efficiency are wiped out by the rising demands and economic activity! Finally, in the year 2010, the per capita energy consumption dropped. The obvious reason for this is the recession that started in the year 2008.

In summary, while better energy efficiency for all devices is a very desirable goal to pursue, it is not a magic wand that will reduce the consumption of energy. In a society that is obsessed with economic growth, in a world that is still growing in population, and in cultures where splendor and wealth are endlessly pursued, energy consumption is only going to increase.

Future Scenarios

Now that various aspects related to energy resources have been analyzed, it will be a logical conclusion to discuss the future for energy resources. To a great extent, the future for energy resources also dictates the future of our civilization. Making a firm prediction for the future is almost an impossible task for energy resources. Firstly, it is impossible to predict the scientific breakthroughs that may occur. Scientists and engineers are good problem solvers. Securing a sustainable and growing energy supply that retains the integrity of biosphere is a complex and interesting problem to solve. A solution to such a problem will not only bring accolades from the scientific community but it will also be a great business opportunity that will eclipse some of the most profitable businesses in the recent decades. The challenge is to engineer a solution within constraints imposed by the laws of thermodynamics. It is impossible to predict if such a scientific breakthrough can take place in the next two or three decades.

Another problem in predicting the future is the difficulty in predicting the behavior of our civilization when faced with the problem of shrinking, or at least a plateau of, energy supply. As discussed earlier, such a situation will result in severe, global economic problems. The response to such a situation by nations could well be in the form of wars to secure energy resources. Even within nations, severe economic problems can lead to turmoil. It is impossible to predict the events that will unfold.

Finally, the impact of rising CO_2 levels in the atmosphere is difficult to predict. A catastrophic climate change will not only impact the economy and the energy use, but also the very structure of our civilization. However, the science related to climate change is still evolving and it is not possible to make firm predictions for climate change.

Hence, instead of making a prediction, it will be prudent to discuss the possible scenarios for the future. The worst-case, best-case, and the most-likely scenarios are presented as a conclusion to this book. These are nothing but educated guesses and it will be intriguing to see the events unfold in the next two or three decades.

Worst-Case Scenario

The worst-case scenario is based on an interlinked occurrence of several events in the next couple of decades:

- Peak oil becomes a reality for the conventional, cheap crude oil. Production declines from most oil producing fields in the Persian Gulf. The non-conventional oil sources; viz. shale oil and tar sands, are unable to ramp up production to match the decline. The reason could either be the cost or that the environmental damage associated with harnessing of such resources becomes unacceptable.

- Decline in oil production results in rising oil prices. This price rise results in an increasing demand for natural gas, for which the price rises as well. Yet, natural gas is unable to replace crude oil derivatives and prices of both remain high. While shale gas does help, the short lifespan of shale gas wells and the requirement to keep drilling a large number of wells leads to a limit on the shale gas output.

- Increase in the price of natural gas results in a rise in electricity prices. There is a rise in demand for coal as a cheap source for electricity production.

- Nuclear energy fails to take off in India and China due to a serious accident in a FBR. In the OECD nations, no new power plants gets built due to public opposition to nuclear energy. Plants built in the 1970s and 80s near their end of life and start getting decommissioned.

- In absence of any technological breakthrough, renewable sources of energy remain expensive, and as a result, remain a small part of the total energy supply. Hydroelectricity still supplies a useful amount of electricity but very little capacity addition takes place as most of the resource-base gets harnessed.

- The task of fighting global warming becomes very urgent due to retreating glaciers, water shortages, rising sea levels, and contracting arctic ice-cap. There is a fear that certain tipping points may be breached very soon, leading to an irreversible, rapid, and catastrophic climate change. It becomes necessary to put a hard limit on CO_2 emissions by every nation.

- The double whammy of energy supply crunch and climate change severely impact the world economy. A worldwide economic recession leads to closing of businesses and layoffs. The banking system is hard hit due to a large scale default on loan payments by corporates and individuals. The standard of living is lowered across the world which reduces the demand for energy. As energy prices moderate due to reduced demand, world economy stumbles back into growth mode. However, improving economy once again leads to high energy prices that bring back the recession. This cycle continues until the standard of living is lowered significantly in most nations and the gains in human development over the last century are erased. There are stresses during these cycles that lead to anarchy and wars. It may be argued that the great recession that started in the year 2008 was the first phase of these cycles.

This is indeed a very depressing scenario. Personally, I'm more hopeful and would like to present the best-case scenario before coming to the most likely outcome.

Best-Case Scenario

The best-case scenario is not based on elusive technologies such as nuclear fusion or fast breeder reactors. Neither does it assume a breakthrough in photovoltaics or other renewable technologies. While nanotechnology does offer some hope of improving efficiency and reducing the cost of photovoltaic panels, it cannot magically improve the energy density of photovoltaic power. A low energy density implies that a system that delivers a significant amount of energy will need to be spread out over a large area. Any system that is spread out over a large cannot avoid the associated costs, whether they are land costs or the cost of large number of resources that a large installation will necessitate. So, the best case scenario is based on small improvements in a number of aspects instead of the magic wand of technological breakthroughs. Here is a list of correlated events and assumptions for the best case scenario:

- The linkage between human development and energy consumption is fairly strong. An HDI of 0.8 requires a per capita energy consumption of at least 100 GJ/year. What if the HDI of 0.8 could be achieved at just 50 GJ/capita/year, and an HDI of 0.9 required only 100 GJ/capita/year? There is nothing inherently impossible about these targets. For instance, in the year 2010, UK consumed only about 140 GJ/capita/year while enjoying a high HDI. However, the prerequisite for such a change is the rational use and equitable distribution of energy and resources. For instance, if energy consumption in the US reduces from 300 GJ/capita/year to 100 GJ/capita/year, the saved energy, at least in theory, can increase the energy availability in India to 70 GJ/capita/year. Such an increase will be sufficient to raise the HDI in India to more than 0.8. Even within a country such as India, the difference in energy use by the affluent and the poor is nothing but unjust. In the best-case scenario, the world will switch to a rational use of energy and other resources, along with an equitable distribution.

- A rational use of resources is easier said than done. Achieving such a change in the next two to three decades, and managing the inevitable economic impact due to reduced consumption, will be a tough challenge. A detailed discussion on this topic is perhaps the subject of another book and I won't attempt to venture into any details in this book. It is sufficient to say that such a change will require the curbing of certain very basic instincts that evolution has ingrained in us.

- Change to a rational and equitable use of energy over the next two to three decades will result in moderation in the demand for fossil fuels. Availability of a stable supply of crude oil over the next two or three decades at the current (2012), inflation adjusted, price of about $100 per barrel, is a necessity. In other words, even though the conventional crude oil supply is unlikely to increase much from the current (2012) level, shale oil and other non-conventional sources should make it possible to maintain the supply as global consumption reduces.

- A better infrastructure is necessary to supply large quantities of natural gas from Qatar, Turkmenistan, and Iran, to India and China. Such a supply can reduce the use of coal for electricity generation in these nations. The natural gas infrastructure not only includes LNG terminals and ships but also international pipelines.

- Shale gas resources in the US may prove to be large and accessible to maintain supply at a reasonable price for the next two to three decades. Assuming that energy-use in the US reduces significantly over the next two decades, the demand for natural gas will be lower. The large deposits of shale gas will make it possible to supply natural gas for a long time.

- Research in nanotechnology may result in marginal improvement in the efficiency of photovoltaic panels while reducing the manufacturing costs. Super-capacitors, which is a new technology for energy storage, may become available at a price comparable with that of lead-acid and NiMH batteries. The net result of these improvements is the increase in the capacity factor for photovoltaic electricity. It may become practically possible to achieve a 100% capacity factor at a cost less than that of systems available today without battery storage. With these improvements, photovoltaic electricity will become the most democratic energy resource.

- For solar thermal electricity, molten salt storage technology can be field-proven in the next few years. Improvements in efficiency and capacity factor, with a reduction in costs, can make solar thermal electric power available at a cost comparable with that of photovoltaic electricity. It may be possible to build large, centralized systems which can satisfy a bulk of electricity demand.

- While there is no solution for the problem of low energy density for photovoltaic and solar thermal electric systems, significant improvement in capacity factor and lower costs can make it possible to deploy these systems widely to tap the huge resource.

- Nuclear energy and hydroelectricity can provide a stopgap arrangement, until photovoltaic and solar thermal electric systems get deployed widely. Of course, hydroelectricity will continue to be a useful resource due to its low cost, dispatchability, and the large resource-base.

- Climate change due to the increasing CO_2 concentration may, hopefully, proceed at a slow pace, and no tipping points will be crossed during this century. This allows for sufficient time to the mankind to take corrective actions; viz. significant reduction in the consumption of fossil fuels to become a carbon-neutral society; stabilizing, and possibly reducing the population; changing the monetary and economic systems that require persistent economic growth; achieving an equitable distribution of resources.

Most-Likely Scenario

The best case and worst case scenarios are clearly very extreme. I am not optimistic enough to predict that the civilization will adjust to a rational and equitable use of resources in a matter of two or three decades on a global scale. Neither am I so pessimistic to accept that the civilization will collapse as a result of an energy crisis. The most likely scenario will be something in the middle. No one can predict the future and calling a certain scenario to be the *most likely* is a matter of personal opinion. However, there cannot be any doubt about one conclusion: On the time scale of a civilization, the era of high energy-use with the aid of fossil fuels is going to be a rather short one. The civilization may last for several millenniums. The fossil fuels, at the current rate of use, are not going to last very long on such a time-scale. The cornucopian theory does not work for energy resources beyond a certain limit, due to the second law of thermodynamics. As a result, the net fossil-fuel supply is going to be limited, and the civilization will need to face the depletion of fossil fuels at a certain point in time.

There is a trend over the past few years that energy consumption in the developed nations is no longer increasing while that in the developing world is rising steadily. The TPE consumption of 5528 MTOE by the OECD nations in the year 2011 was less than the consumption of 5669 MTOE in the year 2005. During the same period, TPE consumption in the non-OECD nations increased from 5086 MTOE to 6747 MTOE [BP 2012]. There are several reasons for this:

- Economic slowdown in the OECD nations, while developing nations continued on the trajectory of growth.
- Improvements in energy efficiency, especially in the developed nations.
- Almost no growth in population in the developed world, while population continues to grow in the developing nations.

This data indicates an important trend. The developed nations have probably reached the limit for economic growth and hence the energy consumption by these nations has peaked. The developing nations are continuing to grow and consume more energy as millions come out of poverty, resulting in a robust internal consumption. A direct result of this will be a continued high demand for energy resources which will keep the energy prices high. At the same time, there is a limit to which the supply of crude oil and natural gas can increase. This situation will lead to a rising consumption of coal in the developing nations, resulting in higher CO_2 emissions and pollution. Nuclear energy will probably stagnate and the renewable sources of energy will take a very long time to supply a significant portion of energy consumption. This situation will lead to high energy prices, economic stagnation in the developed world, and low growth in the developing nations. The impact of economic hardships will force the developing nations to change their models of development. Instead of blindly following the path of the OECD nations, hopefully, sense will prevail. There will be a realization that development does not imply creating another USA, as far as resource consumption is concerned. The basic human necessities of water, food, energy, education, health-care, affordable housing, roads, public transport infrastructure, and functioning democratic institutions will get the priority in the development model. With such a change, it may be possible to achieve good human development while consuming only a reasonable amount of energy. I am hopeful that even the developed nations will adjust to the new reality and change to a rational and sustainable use of energy. Such a change may take several decades and this time period will certainly be a difficult one. Hopefully, the global warming will not progress rapidly, allowing time for all the changes to take place. In summary, the most-like scenario in my opinion is that a transition will take place over the next 50 years to a society that uses significantly less energy and resources. This transition will be difficult but exciting at the same time.

I hope that this book has given you the information and direction to gain an understanding of the world of energy resources. I encourage you to write and keep record of your version of the most-likely scenario. Thinking about this will certainly influence the decisions that you take about all aspects of life, whether they are about investment, career, family, pleasure, or recreation. Without any intention of preaching, I'll quote Thomas Jefferson to end this book:

It is not the wealth or splendor, but tranquility and occupation which give happiness.

REFERENCES

Barbose G., et al. (2011). *Tracking the Sun IV.* Environmental Energy Technologies Division, Lawrence Berkley National Laboratory, US Department of Energy. Retrieved Oct 20, 2012, from *<http://eetd.lbl.gov/ea/emp/re-pubs.html>*

BEE: <http://www.beeindia.in>. India: Bureau of Energy Efficiency

Biello, D. (2009, February 18). *How to Use Solar Energy at Night.* Scientific American. Retrieved Apr 22, 2011, from, *<http://www.scientificamerican.com/article.cfm?id=how-to-use-solar-energy-at-night>*

Boccard N. (2009). Capacity Factor of Wind Power, Realized Values vs. Estimates. *Energy Policy* 37:2679-2688. Retrieved Aug 24, 2012, from, *<http://www.elsevier.com/locate/enpol>*.

Boxell, J. (2012, July 16): Areva's Atomic Reactor Faces Further Delays, *The Financial Times*, Retrieved Jan 17, 2013, from, *<http://www.ft.com>*.

BP (2010, June). *The BP Statistical Review of World Energy.* Retrieved Jan 24, 2011, from *<http://www.bp.com/statisticalreview>*. London: BP p.l.c.

BP (2011, June). *The BP Statistical Review of World Energy.* Retrieved Nov 4, 2011, from, *<http://www.bp.com/statisticalreview>*. London: BP p.l.c.

BP (2012, June). *The BP Statistical Review of World Energy.* Retrieved June 21, 2012, from, *<http://www.bp.com/statisticalreview>*. London: BP p.l.c.

BrightSource (2012). *Ivanpah Project Facts.* Retrieved Nov 8, 2012, from, *<http://www.brightsourceenergy.com>*. Oakland, CA: BrightSource Energy.

Chopra, A. (2011, Mar 28). A Perspective on the Nuclear Uproar in India. *Forbes India.* Retrieved Jan 17, 2013, from, <*http://forbsindia.com*>.

CJP (2012). *Jatropha Biodiesel Economics.* Retrieved on Dec 10, 2012, from, <*http://www.jatrophaworld.org*>. Rajasthan, India: The Center for Jatropha Promotion.

Cochran, T., et al. (2010, February): *Fast Breeder Reactor Programs: History and Status.* Retrieved January 24, 2013, from, <*http://www.fissilematerials.org*>. Princeton, NJ: International Panel on Fissile Materials, Princeton University.

Cree (n.d.). *Xlamp XP-E LED Datasheet.* Retrieved May 14, 2012, from, <*http://www.cree.com/led-components-and-modules/products/xlamp/discrete-nondirectional/xlamp-xpe-hew*>. Durham, NC: Cree Inc.

CWET (2012). *Estimation of Installable Wind Power Potential at 80 m level in* India. Retrieved Aug 26, 2012, from, <*http://www.cwet.tn.nic.in*>. Chennai, India: The Center for Wind Energy Technology,

ENERGY STAR (2010). *Overview of 2010 Achievements,* Retrieved May 14, 2012, from, <*http://www.energystar.gov*>.

EIA. <http://eia.doe.gov>. Washington DC: US Energy Information Administration, US Department of Energy.

EIA (2011, October). *Annual Energy Review 2010.* DOE/EIA-0384(2010). Washington DC: US Energy Information Administration, US Department of Energy. <*http://eia.doe.gov/aer*>

ET (2012). RIL cuts proven gas reserves estimates for KG-D6 block. *Economic Times,* May 9, 2012. <*http://articles.economictimes.indiatimes.com/2012-05-09/news/31641932_1_ril-cuts-kg-d6-shale-gas*>.

Gagnon, L. (2005, July). *Comparing Energy Options.* 2005G185-A. Retrieved August 13, 2012, from, <*http://www.hydroquebec.com/sustainable-development*>. Canada: Hydro-Quebec.

Grand Coulee (2012, March). *Grand Coulee Dam Statistics and Facts.* Retrieved Aug 1, 2012, from, <*http://www.usbr.gov*>. Boise, Idaho: Bureau of Reclamation, US Department of the Interior.

GWEC (2011, April). *Indian Wind Energy Outlook 2011.* Retrieved August 24, 2012, from, <*http://www.gwec.net*>. Brussels, Belgium: Global Wind Energy Council.

ICOLD (2012). *Role of Dams*. Paris: International Commission on Large Dams. <*http://www.icold-cigb.net/*>

IEA (2011). *Key World Energy Statistics*. Retrieved Jan 23, 2012, from, <*http://www.iea.org*>. Paris: International Energy Agency.

Jayakumar, P. (2012, Apr 2). The Power Industry's Ultra Mega Problems. *Business World*, vol. 31-46, p. 36-41.

Johns Hopkins University (2011, February 7). Better turbine spacing for large wind farms. *ScienceDaily*. Retrieved August 20, 2012, from <*http://www.sciencedaily.com-/releases/2011/01/110120111332.htm*>

Kakodkar, A. (2009, July). Preparing for Our Technology Future. *Physics News,-Bulletin of IPA*, vol. 39, p. 5-14.

Kakodkar, A. (2011, January 5). A Firm Step Towards Energy Independence (article in Marathi), *Pune Sakal*, January 5, 2011.

Koyna. *The Koyna Hydroelectric Project*. Mumbai, India: Water Resources Dept, Govt of Maharashtra. <*http://www.koynaproject.org*>

Kyocera (n.d.). *Kyocera KD215GX-LPU Solar Panel Specification*. Retrieved Feb 28, 2011, from, <*http://www.kyocerasolar.com*>. Kyocera Solar Inc.

Lengyel, G. (2007). *Department of Defense Energy Strategy, Teaching an Old Dog New Tricks*, Washington, D.C.: The Brookings Institution.

Makhijani, A. (2010). *The Mythology and Messy Realty of Nuclear Fuel Reprocessing*. Retrieved Jan 7, 2011, from, <*http://www.ieer.org/reports/reprocessing2010.pdf*>. Takoma Park, MD: Institute for Energy and Environmental Research.

MEDA (2009). *Wind Power Generation for 2008-09*. Data Retrieved May 20, 2011, from, <*http://www.mahaurja.com/PG_WE_Overview.html*>. Mumbai, India: Maharashtra Energy Development Agency, Government of Maharashtra.

MIT (2011). *The Future of Natural Gas, an MIT Interdisciplinary Study*. Retrieved Mar 1, 2012, from, <*http://mitei.mit.edu/publications/reports-studies*>. Cambridge, MA: MIT.

NASA EOS. *Surface Meteorology and Solar Energy*. Atmospheric Science Data Center, NASA, USA. <*http://eosweb.larc.nasa.gov/sse/*>

NHPC. National Hydroelectric Power Corporation of India Ltd, <http://www.nhpcindia.com>

NREL (2011). *Estimates of Windy Land Area and Wind Energy Potential for Areas >= 30% Capacity Factor at 80m.* Retrieved Aug 26, 2012, from, <http://www.nrel.gov/wind>. *Golden, CO:* National Renewable Energy Laboratory.

Ramana, M. (2009). India and Fast Breeder Reactors. *Science and Global Security*, 17:54-67, 2009.

Sardar Sarovar. Sardar Sarovar Narmada Nigam Ltd. <http://www.sardarsarovardam.org>

Solar India (n.d.). *Towards Building Solar India.* Retrieved Aug 7, 2011, from, <http://india.gov.in/allimpfrms/alldocs/15657.pdf>. New Delhi: Jawaharlal Nehru National Solar Mission, Govt of India.

Smil, V. (2005). *Energy at the Cross-roads.* Cambridge, MA: The MIT Press.

SunPower (2011). *E20/327 Solar Panel Specification.* Retrieved October 19, 2012, from, <http://www.sunpowercorp.com>. SunPower Inc.

Suzlon (2011). *S9X Wind Turbine Technical Specification*, Retrieved August 21, 2012, from, <http://www.suzlon.com>. Suzlon Energy Ltd.

Tehri. *The Tehri Dam & Hydro Electric Project.* THDC India Ltd. <http://thdc.gov.in>

TOI (2012). India to sign pact next week on TAPI gas pipeline, *The Times of India*, May 17, 2012. Retrieved June 29, 2012, from, <http://timesofindia.indiatimes.com>

UN Data: United Nations Database. <http://data.un.org>

USGS (2011, August). *Mean Shale Gas Resources.* Retrieved March 2, 2012, from, <http://energy.usgs.gov>.

Wiser R. and Bolinger M. (2012, August). *2011 Wind Technologies Market Report.* Retrieved August 16, 2012, from, <http://www1.eere.energy.gov/wind>. Wind and Water Power Program, Lawrence Berkley National Laboratory, US Department of Energy.

WNA (2010). *Supply of Uranium.* Retrieved Jan 22, 2011, from, <http://world-nuclear.org/info/Nuclear-Fuel-Cycle/Uranium-Resources/Supply-of-Uranium>. World Nuclear Association.

WNA (2012) *Fast Neutron Reactors*. Retrieved Jan 15, 2013, from, <http://*www.world-nuclear.org/info/Current-and-Future-Generation/Fast-Neutron-Reactors*>. World Nuclear Association.

WNA (2013). *Fukushima Accident 2011*. Retrieved Jan 28, 2013, from, <http://*www.world-nuclear.org/info/Safety-and-Security/Safety-of-Plants/Fukushima-Accident-2011*>. World Nuclear Association.

WNA (2013a). *World Nuclear Power Reactors and Uranium Requirements*. Retrieved Jan 30, 2013, from, <*http://www.world-nuclear.org/info/Facts-and-Figures/World-Nuclear-Power-Reactors-and-Uranium-Requirements*>. World Nuclear Association.

WNADB. *Reactor Database at the World Nuclear Association*. <*http://www.world-nuclear.org*>

ACRONYMS AND GLOSSARY

CNG: Compressed Natural Gas is essentially Methane (CH_4).

Capacity Factor: Actual electrical energy output of a power station as compared to the maximum generation capacity. The capacity factor is normally measured over a long interval that can range from few weeks to a year.

CERN: *Conseil Européen pour la Recherche Nucléaire or,* The European Organization for Nuclear Research, headquartered in Geneva, Switzerland. This is an international organization where scientists from all over the world collaborate to conduct research in nuclear physics.

CFL: Compact Fluorescent Lamp

DNI: Direct Normal Insolation is the insolation received while tracking the movement of sun in the sky.

EMF: Electromotive Force

Energy Intensity of Economy: Energy intensity is the average amount of primary energy consumed per unit economic output.

EROEI: Energy Returned on Energy Invested. Please refer page 21 for a detailed explanation.

EUR: Expected Ultimate Recovery. Total output over the lifetime of an oil or gas well.

FBR: Fast Breeder Reactor. A nuclear reactor that makes use of *fast* neutrons to breed more fuel than it consumes.

GDP: Gross Domestic Product

GHG Emissions: Greenhouse Gas Emissions. Carbon dioxide (CO_2) and Methane (CH_4) are the two major greenhouse gases.

GNI: Gross National Income

HDI: Human Development Index. A number published by the UN that attempts to measure the overall development of a nation.

Heavy Water: D_2O. Water that consists of deuterium instead of hydrogen. Deuterium is an isotope of hydrogen that contains one neutron in addition to the proton in the nucleus.

IAEA: International Atomic Energy Agency. It is an independent international organization within the United Nations family. The IAEA website states that it works with its member states and multiple partners worldwide to promote safe, secure, and peaceful nuclear technologies.

ICE: Internal Combustion Engine. Almost all automobiles use an ICE as the prime mover that converts chemical energy in the fuel to mechanical energy.

LED: Light Emitting Diode. An electronic device that converts electrical energy to light. Modern LED lights achieve conversion efficiencies higher than even the fluorescent lamps.

LNG: Liquefied Natural Gas. Natural gas, which mostly contains methane (CH_4), liquefies when cooled to -162° Celsius. The reduction in volume that takes place due to liquefaction results in a high energy density, making transportation over long distance economical.

LPG: Liquefied Petroleum Gas. This is a mixture of propane (C_3H_8) and butane (C_4H_{10}), both of which are hydrocarbons. LPG is primarily used as a fuel. It is stored in steel cylinders at high pressure to maintain in a liquid form.

LWR: Light Water Reactor. A type of nuclear reactor that makes use of light water (H_2O) as moderator and coolant.

MTOE: Million TOE. See, TOE.

MW/MWe/GW/GWe: MWe (GWe) specifically identifies electrical power while MW (GW) is a general unit for power. In a thermal power station with an efficiency of

38%, a thermal power input of 100 MW is necessary to generate an electrical output of 38 MWe.

NASA: National Aeronautics and Space Administration, USA

NGLs: Natural Gas Liquids. These are the heavier hydrocarbons found along with methane in a natural gas well. They mostly consist of ethane, propane, butane, and iso-butane.

NREL: National Renewable Energy Laboratory, Department of Energy, USA

OECD Countries: Organization for Economic Cooperation and Development. This is a group of nations that consists of following members: Australia, Austria, Belgium, Canada, Chile, Czech Republic, Denmark, Estonia, Finland, France, Germany, Greece, Hungary, Iceland, Ireland, Israel, Italy, Japan, Korea, Luxembourg, Mexico, Netherlands, New Zealand, Norway, Poland, Portugal, Slovak Republic, Slovenia, Spain, Sweden, Switzerland, Turkey, United Kingdom, United States. Most of these countries are highly developed and the group accounts for about half of the world TPE consumption.

OPEC: Organization of Petroleum Exporting Countries. The members of OPEC are: Algeria, Angola, Ecuador, Iran, Iraq, Kuwait, Libya, Nigeria, Qatar, Saudi Arabia, UAE, Venezuela.

PHWR: Pressurized Heavy Water Reactor. A type of nuclear reactor that makes use of heavy water (D_2O) as moderator and coolant.

PLF: Plant Load Factor. See Capacity Factor.

PPP: Purchasing Power Parity. This is the adjustment to GNI/GDP that takes into account the real purchasing power of national currency and not just the exchange rate with respect to the US dollar (or any other reference currency).

PWR: Pressurized Water Reactor. This is an LWR that uses water under high pressure in the reactor core to prevent boiling.

R/P Ratio: The ratio of Reserves to Production. This ratio is normally used for non-renewable resources such as fossil fuels. This ratio indicates the number of years for which the reserves are expected to last, provided there are no additions to the reserves and production remains unchanged.

SI: *Système international d'unités* (French). The international system of units for the measurement of physical quantities.

Synchronus Inverter: An electronic device that converts a DC electric supply into an AC supply such that the AC output is in perfect synchronization with the supply from electric grid. Perfect synchronization implies a perfect match in frequency, phase, amplitude, and waveform. A synchronus inverter makes it possible to connect a solar panel to the electric grid.

TOE: Tonne Oil Equivalent. This is the energy content of 1 tonne of crude oil, which approximates to 42 gigajoules (GJ). It is a common practice to make use of TOE and MTOE as units for energy.

TPE: Total Primary Energy.

USGS: United States Geographical Survey.

ABOUT THE AUTHOR

Aniruddha B. Joshi is an engineer by profession. He earned his master's degree in engineering from the Indian Institute of Technology, Powai, Mumbai. Aniruddha has US patents to his credit in the field of computer networking. He has been working as a hobbyist in the field of renewable sources of energy since 2004. Aniruddha lives in Pune, India. He can be reached via email to: abjoshi@gmail.com

Index

www.ingramcontent.com/pod-product-compliance
Lightning Source LLC
Chambersburg PA
CBHW081309170526
45166CB00011B/3459